THE GENESIS MISSION

THE GENESIS MISSION

Edited by

CHRISTOPHER T. RUSSELL
University of California, California, U.S.A.

Reprinted from *Space Science Reviews*, Volume 105, Nos. 3–4, 2003

KLUWER ACADEMIC PUBLISHERS
DORDRECHT / BOSTON / LONDON

A.C.I.P. Catalogue record for this book is available from the Library of Congress

ISBN: 1-4020-1125-3

Published by Kluwer Academic Publishers,
P.O. Box 990, 3300 AZ Dordrecht, The Netherlands

Sold and distributed in North, Central and South America
by Kluwer Academic Publishers,
101 Philip Drive, Norwell, MA 02061, U.S.A.

In all other countries, sold and distributed
by Kluwer Academic Publishers,
P.O. Box 322, 3300 AH Dordrecht, The Netherlands

Printed on acid-free paper

Printed in the Netherlands

SPACE SCIENCE REVIEWS / *Vol. 102 Nos. 1–2 2002*

TABLE OF CONTENTS

Foreword

Genesis is the fifth mission in the Discovery series of cost-capped planetary missions, and the second to attempt a sample return to Earth. Because of its short interplanetary trajectory, it will be the first to return such a sample. Genesis' sample is solar matter, not planetary per se, but, since we believe that the Sun and the planets condensed out of a homogeneous nebula, the Sun's composition is key to interpreting meteoritic evidence on the origin and evolution of the solar system. Back on Earth Genesis' samples will be analyzed by precision laboratory instruments.

This special issue of Space Science Reviews described the objectives of the Genesis mission, the collector materials used to capture the solar wind material, the solar wind concentrator and simulations of its performance, the plasma instrumentation and how it is used to classify the solar wind flow.

The editor wishes to thank the referees of this volume whose suggestions have helped the authors produce a most useful document allowing the many scientists who will use these data to understand the circumstances under which they were acquired.

C. T. Russell
University of California
Los Angeles, CA
August 16, 2002

THE GENESIS DISCOVERY MISSION: RETURN OF SOLAR MATTER TO EARTH

D. S. BURNETT[1], B. L. BARRACLOUGH[2], R. BENNETT[3], M. NEUGEBAUER[3], L. P. OLDHAM[4], C. N. SASAKI[3], D. SEVILLA[3], N. SMITH[4], E. STANSBERY[5], D. SWEETNAM[3] and R. C. WIENS[2]

[1]*Division of Geological and Planetary Sciences, California Institute of Technology, Pasadena, CA 91125, U.S.A.*

[2]*Space and Atmospheric Sciences, MS D466, Los Alamos National Laboratory, Los Alamos, NM 87545, U.S.A.*

[3]*Jet Propulsion Laboratory, Pasadena, CA 91109, U.S.A.*

[4]*Lockheed-Martin Astronautics, Denver, CO 80201, U.S.A.*

(Author for correspondence, email: burnett@gps.caltech.edu)

Received 25 March 2002; Accepted in final form 26 August 2002

Abstract. The Genesis Discovery mission will return samples of solar matter for analysis of isotopic and elemental compositions in terrestrial laboratories. This is accomplished by exposing ultra-pure materials to the solar wind at the L1 Lagrangian point and returning the materials to Earth. Solar wind collection will continue until April 2004 with Earth return in Sept. 2004. The general science objectives of Genesis are to (1) to obtain solar isotopic abundances to the level of precision required for the interpretation of planetary science data, (2) to significantly improve knowledge of solar elemental abundances, (3) to measure the composition of the different solar wind regimes, and (4) to provide a reservoir of solar matter to serve the needs of planetary science in the 21st century. The Genesis flight system is a sun-pointed spinner, consisting of a spacecraft deck and a sample return capsule (SRC). The SRC houses a canister which contains the collector materials. The lid of the SRC and a cover to the canister were opened to begin solar wind collection on November 30, 2001. To obtain samples of O and N ions of higher fluence relative to background levels in the target materials, an electrostatic mirror ('concentrator') is used which focuses the incoming ions over a diameter of about 20 cm onto a 6 cm diameter set of target materials. Solar wind electron and ion monitors (electrostatic analyzers) determine the solar wind regime present at the spacecraft and control the deployment of separate arrays of collector materials to provide the independent regime samples.

1. Introduction

Genesis is the fifth in NASA's low-cost line of missions called Discovery. The overall purpose of the mission is to collect samples of solar wind and return them to earth. Following launch on August 8, 2001, the Genesis mission began collecting samples of solar matter on November 30, 2001. Exposure of collector materials will be for 27 months near the L1 Sun–Earth libration point. The overall mission trajectory is shown in Figure 1. These samples will return to Earth for laboratory isotopic and chemical analyses on Earth on September 8, 2004. At that time the real Genesis science mission will begin. This paper summarizes the pre-return sci-

Space Science Reviews **105:** 509–534, 2003.

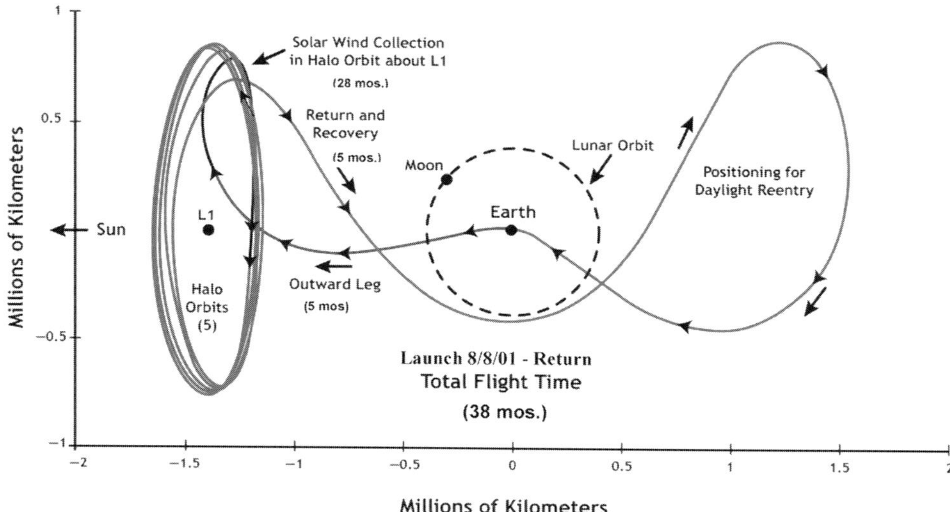

Figure 1. The Genesis spacecraft trajectory from Earth to L1 and return. The Earth-L1 distance is about 10^6 km. The lunar orbit is shown for scale. Arrows indicate the outbound and return trajectories. There are 5 'halo' orbits about L1. The large loop behind the Earth on the return trajectory positions the spacecraft for daylight re-entry.

ence rationale and mission technical approach to provide context to the following important and substantial instrument papers.

2. Science Background

The science goals of NASA are to understand the formation, evolution, and present state of the solar system, the galaxy, and the universe. Most planetary missions investigate the present state of planetary objects. By, in effect, going back in time, Genesis addresses questions about the materials and processes involved in the origin of the solar system by providing precise knowledge of solar isotopic and elemental compositions.

The context for the interpretation of Genesis data is a wide spread consensus on a 'standard model' for the origins of planetary materials. This model assumes that (1) planetary materials and objects arose from a compositionally homogeneous solar nebula. (2) With the exception of D and the Li isotopes, the elemental composition of the nebula is preserved in the solar photosphere and (3) With the exception of part per thousand or less mass dependent isotopic fractionations due to physicochemical processes, isotopic compositions are the same in all parts of the solar system and thus equal to terrestrial values. Genesis data can be used to refine the standard model, but more importantly, test its assumptions.

For Genesis, the solar wind is just a convenient source of solar matter readily available outside the terrestrial magnetosphere. Solar wind ions have velocities in the well-understood ion implantation regime and are essentially quantitatively retained upon striking passive collectors. (Small backscattering loss corrections can be accurately made.)

Solar wind collection was demonstrated by the highly successful Apollo solar wind foil experiments (Geiss *et al.*, 1972). The Apollo foils were only sufficiently pure for the study of He, Ne, and ^{36}Ar. With 100-times longer exposure and, especially, with purer collector materials, the goal of Genesis is to provide precise solar isotopic compositions and greatly improved solar elemental composition for most of the Periodic Table.

3. General Science Objectives

Analysis of the collector materials will give precise data on the chemical and isotopic composition of the solar wind. By returning samples of solar matter to Earth, the Genesis mission will provide:
(1) Isotopic abundances of sufficient precision to address planetary science problems.
(2) Major improvement in our knowledge of the average chemical composition of the solar system.
(3) Independent compositional data on 3 different kinds of solar wind (we refer to these as 'regimes'.)
(4) A reservoir of solar material to be used in conjunction with advanced analytical techniques available to 21st century scientists.

Items (1)–(3) will be discussed further below. The Sample Allocation and Curation section summarizes how we meet Objective (4).

3.1. Solar isotopic compositions

The standard model predicts the same isotopic compositions for all solar system materials; however, measurable heterogeneities exist in the isotopic compositions of some, but not all, elements among various planetary materials. Thus, solar isotopic compositions should be the reference point for comparisons with planetary matter. The specific cases of O and N will be discussed below.

Any measurable isotopic difference is of great importance because, in many cases, these cannot be explained by the same chemical and physical processes which are usually invoked to explain elemental differences (e.g., differences in temperature). It is independently known that the solar system was formed from a wide variety of stellar materials produced over the 5–10 billion years of galactic history prior to the formation of the solar system and that these materials had a great diversity of isotopic compositions. The isotopic compositions of solar matter thus define the average of these diverse inputs and thus represent an anchor point for the interpretations of the isotopic differences among planetary materials. Isotopic variations are expected given that the observed isotopic compositions are the result of mixing and that the mixing may not have been complete. However, there is no *a-priori* constraint on the magnitude of variations. For many elements, the variations among presently-available extraterrrestrial materials are less than parts in 10^3 to 10^4, indicating rather thorough mixing.

Solar composition is important for astrophysics and solar physics, but planetary science requires greater elemental coverage and much higher levels of precision, especially for isotopes. For example, most theories of stellar nucleosynthesis are considered successful if solar system isotope ratios are reproduced within a factor of 2. By contrast, isotopic measurements of terrestrial, lunar, Martian, and meteoritic materials typically deal with 0.1% and smaller differences. In atmospheric modeling, differences of 1 percent or less are crucial for, e.g., $^{38}Ar/^{36}Ar$ and the Xe isotopes. As discussed below, O and N are important exceptions, showing a range of around 5% (for both $^{18}O/^{16}O$ and $^{17}O/^{16}O$) and 60% for $^{15}N/^{14}N$. For comparison the most precise spacecraft measurement of $^{18}O/^{16}O$ has uncertainties of $\pm20\%$ and ^{17}O cannot be detected with present instruments. (Wimmer-Schweingruber *et al.*, 2001).

3.2. Elemental abundances

The observed diversity in solar system objects is chemical in origin. Quantitatively, diversity can be defined as the difference in planetary material composition from solar composition, illustrating the importance of solar elemental abundances. The present best source of solar abundances comes from analysis of photospheric absorption lines in the solar spectrum. A small number of elements have quoted errors of $\pm10\%$ (one sigma), but overall there are large uncertainties in these abundances and a significant number of elements cannot be measured. Thus, compilations of 'solar' abundances for non-volatile elements are currently based on analyses of carbonaceous (CI) chondrite meteorites. The limitations to this have been discussed (Burnett *et al.*, 1989; McSween, 1993). Solar abundances should be based on solar data. For example, it is likely that new CI-like meteorites will eventually become available which will have slightly different abundances than known CI meteorites, presenting a major challenge to how well we think we know solar abundances. If solar composition is based on solar data, we are immune to such periodic perturbations. The best hope for major improvement in knowledge of solar abundances is the solar wind. The Genesis goal is to improve knowledge of solar elemental abundances by at least a factor of 3 for each element.

3.3. Solar Wind Regimes

It is well established that the solar wind is accelerated by more than one mechanism, leading to three major solar wind regimes: high-speed streams from coronal holes, low-speed interstream wind, and the transient wind associated with coronal mass ejections (Neugebauer, 1991; Fisk *et al.*, 1998). Measurements by in-situ solar wind instruments have shown that there is an *elemental* fractionation of matter between the photosphere and the solar wind, which depends on the first ionization time (FIT) and, perhaps to a much lesser extent, on the ion mass and charge (e.g., Marsch *et al.*, 1995; von Steiger *et al.*, 2000). Genesis will collect separate samples for each of these three regimes in order to provide better elemental data on compositional differences among the three regimes (Neugebauer, *et al.*, 2003).

The data for the individual regimes provide a critical test of the accuracy of the corrections of *elemental* abundance data for FIT fractionation effects. We will combine the Genesis solar wind monitor data (Barraclough *et al.*, 2003) with knowledge of the high time resolution systematics of FIT, mass, and charge fractionation patterns obtained from the results of the Ulysses, WIND, SOHO, and ACE missions to model the corrections to be applied to the Genesis sample data to deduce the photospheric elemental composition. It should be emphasized that the observed FIT variations appear only to affect elements with relatively high ionization potential. Present data are consistent with no fractionation relative to photospheric abundances for elements with a first ionization potential less than about 9 eV, which includes most of the Periodic Table and especially all the rock-forming elements which make up the terrestrial planets. For these elements, corrections, if any, appear to be small, but this will be directly assessed by Genesis data. It will not be necessary to *assume* that the corrections are small. Genesis will provide three independent regime data sets which will have different amounts of correction, but yet must give the same corrected photospheric composition. This should provide proof that our derived photospheric elemental abundances are correct.

Although the FIT fractionations are based on elemental properties and thus should not produce isotopic variations, isotopic variations are still possible (e.g., Bochsler, 2000). There is some evidence for transient isotopic variations under different solar wind conditions (Kallenbach *et al.*, 1999; Kallenbach, 2001). Except for $^{3}He/^{4}He$ which shows large transient variations, reported variations are at the precision limits of spacecraft instruments, typically 10–20%, 2 sigma. With the Genesis samples, systematic searches for long term isotopic variations between solar wind regimes can be carried out with much higher precision, 1% (2 sigma) or better for many elements. If variations are found, these will be very important data for solar physics but a complication for the Genesis goals, as corrections will be required to obtain photospheric isotopic compositions. Of course, the solar physics theories developed to account for any regime variations are the basis for correction back to photospheric compositions, illustrating the strong complementarity between Genesis and solar physics goals.

4. Examples of Specific Measurements

Mission science planning, especially the selection and testing of collector materials, was based on a set of prioritized measurement objectives. Our three highest priority objectives are discussed here. The others, along with other Genesis science documents, are discussed at http://www.gps.caltech.edu/genesis/

The highest priority specific objectives are to measure the relative amounts of:

- *O isotopes*, because they provide the basis for understanding observed meteorite variations (e.g., Clayton, 1993; Wiens *et al.*, 1999).
- *N isotopes*, because they are a key reference point in the eventual understanding of large but totally unexplained, N isotopic variations in planetary materials (e.g., Owen *et al.*, 2001).
- *Noble gas isotopes and elements*, because they provide the basis for interpreting the compositions of terrestrial planet atmospheres (e.g., Pepin, 1991).

4.1. O ISOTOPES

Figure 2 is a schematic summary of O isotopic variations in planetary materials, recognizing that all data come from materials residing at 1–3 AU. The scales have a total range of about 8%, and on this scale, as indicated on Figure 2, the error bars on spacecraft or spectroscopic solar data are larger than the figure. Most meteoritic (asteroidal) materials have O isotopic compoistions which lie within a few % of the Earth, Moon, and Mars; however CAI materials (Ca-Al-rich inclusions) from chondritic meteorites show large ^{16}O enrichments. Different models for the origin of the variations predict the location of the solar isotopic composition. One of these (Clayton and Mayeda, 1984) is indicated by the point SM. Other models (e.g., Clayton, 2002) predict that the solar O isotopic composition lies at the most ^{16}O-rich end of the CAI trend. The Genesis precision indicated on Figure 2 can clearly distinguish between these and other models. On a more model-independent basis, a precise measurement of the solar O isotopic composition addresses the fundamental issue of the degree of gas-dust mixing that occurred in the solar nebula for the solid materials that now comprise the inner solar system (Wiens *et al.*, 1999). As they are rocky materials, inner solar system bodies can, in principle, be made from the dust phase of the solar nebula. If isotopic equilibrium between gas and dust was not established prior to dust-gas separation, significant differences could exist between the solar O isotopic composition and any inner solar system material. Operationally, if the solar O isotopic composition measured by Genesis lies close to the Earth-Moon-Mars-asteroid region of Figure 2, it would be indicative of a high degree of gas-dust equilibration. However, if gas-dust equilibration did not occur, the solar O isotopic composition could be significantly displaced from the Earth-Moon-Mars-asteroid region of Figure 2, and this would be our interpretation of a significant displacement.

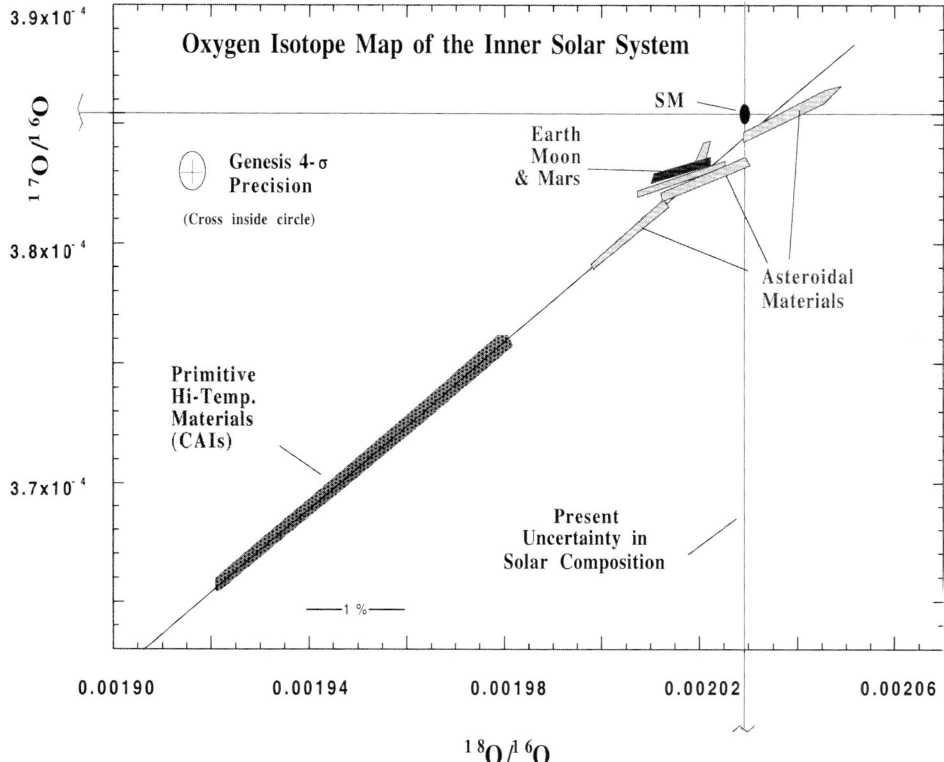

Figure 2. An oversimplified, schematic view of measured O isotopic variations in solar system materials. An absolute value scale is used to indicate the size of the quantities that must be measured. Absolute abundances are not accurately known, but relative values among different materials can be very precisely measured, in many cases down to $\pm 0.1\%$, 2 sigma. On the scale of this figure, the Earth, Moon, and Mars are close, but the range of asteroidal materials, as derived from meteorite analyses, is quite large. The largest variations are found in meteoritic Ca-Al-rich inclusions (CAIs). As indicated, the uncertainties in spacecraft or spectroscopic O isotope measurements are larger than the scale of the figure, whereas the projected Genesis precision, shown as ± 4 sigma for visibility, can resolve various theories as to the location of the solar O isotopic composition. The point labelled SM is a model prediction for the location of the solar isotopic composition.

4.2. N and Noble Gases

These elements are closely coupled. In Figure 3, the $^{15}N/^{14}N$ ratios are expressed as a % difference from the present day terrestrial atmosphere, although this is clearly an arbitrary reference. The Apollo solar wind foils provide a precise value for solar wind $^{20}Ne/^{22}Ne$, with the solar wind Ne isotopic ratio being 38% higher than the value for the terrestrial atmosphere. The difference is of major importance, very likely indicating that the Earth has experienced major atmospheric loss (e.g., Pepin, 1991). If this is true, then differences are also expected between the solar and terrestrial isotopic compositions of N and the other noble gases. Thus,

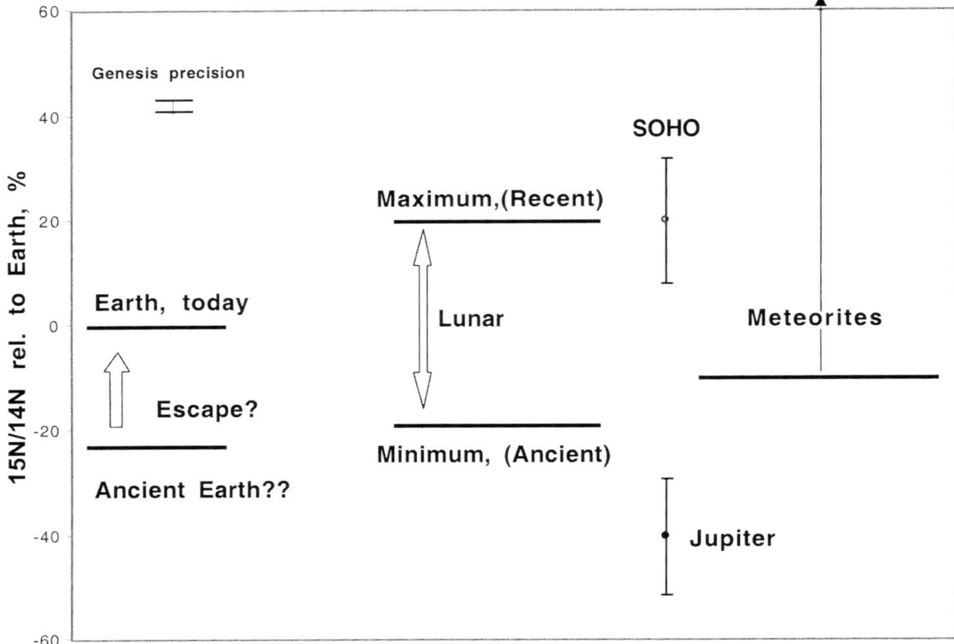

Figure 3. Summary of N isotopic compositions in solar system reservoirs. Meteoritic $^{15}N/^{14}N$ ratios may be affected by the presence of pre-solar materials; however, even excluding these, a 60% range among different materials is observed which is unexplained. The terrestrial atmosphere may be affected by atmospheric loss increasing $^{15}N/^{14}N$. Lunar N isotopic variations, traditionally regarded as reflecting solar wind, have been a long term mystery but there is a suggestion that materials exposed on the lunar surface $> 10^9$ years ago has lower $^{15}N/^{14}N$. There is rough consistency between the ancient terrestrial atmosphere, minimum lunar, and Jovian atmospheric $^{15}N/^{14}N$ that would suggest that the solar $^{15}N/^{14}N$ should be 20–40% lower than the present terrestrial atmosphere, but this is totally contradicted by the SOHO solar wind result which, despite large errors, has a significantly higher $^{15}N/^{14}N$. The indicated Genesis (2 sigma) precision will closely constrain models for the variations.

Genesis will provide major quantitative tests of the atmospheric escape models. Figure 3 is a top level summary indicating that exceptionally large variations in $^{15}N/^{14}N$ characterize different solar system reservoirs For purposes of illustration, a crude estimate of the pre-escape terrestrial atmospheric $^{15}N/^{14}N$, scaled from the $^{22}Ne/^{20}Ne$ differences, is shown as ‘Ancient Earth’ on Figure 3. The exact displacement is highly uncertain; nevertheless, the pre-escape atmospheric $^{15}N/^{14}N$ should be distinctly less than the present day $^{15}N/^{14}N$.

The N and noble gases measured in lunar regolith samples are expected to be of solar origin (e.g., Kerridge, 1989); however, impact heating alteration of lunar surface materials has made the extraction of quantitative solar wind abundances difficult. Over the last 3 decades, an approximate consensus has emerged on ‘solar’ noble gas elemental and isotopic ratios from lunar regolith samples (e.g., Ozima *et al.*, 1998); however, there are many assumptions in arriving at the consensus composition. At the minimum, Genesis data will provide ground truth for the validity of these assumptions. Further, there is strong evidence for time variations in the solar noble gas elemental ratios (Wieler *et al.*, 1996), although not in the isotopic ratios (Ozima *et al.*). Another goal of Genesis is thus to provide present-day solar noble gas elemental and isotopic abundances to help realize the unfulfilled Apollo goal of 'understanding the Sun in time' from the study of lunar samples.

In contrast to the noble gases, there is no consensus on the interpretation of the important and fascinating data for N in lunar regolith samples (e.g., Kerridge, 1989). The lack of understanding of N introduces significant uncertainties in the derived solar noble gas abundances. The N/noble gas ratios are a factor of 10 higher than would be expected from present photospheric abundance estimates. Moreover, N isotopic compositions show a lot of variability with a tendency for samples exposed on the lunar surface in the past to have lower $^{15}N/^{14}N$ than present-day surface soil samples (Figure 3). The ancient-recent trend indicated for the lunar data in Figure 3 is a highly oversimplified, but 'traditional' interpretation in which there has been a 'secular' increase of the solar wind N isotopic composition with time. The prediction for Genesis from the traditional view is that the present-day solar wind $^{15}N/^{14}N$ should be significantly higher than that for the terrestrial atmosphere.

A direct measurement of solar wind $^{15}N/^{14}N$ from SOHO (Kallenbach *et al.*, 1998b) gives a higher value than the terrestrial atmosphere, but not outside a 2 sigma error limit (Figure 3). This is in apparent support of a secular increase in solar wind $^{15}N/^{14}N$; however, there is no known solar physics mechanism to cause such secular variations. Moreover, there are some lunar samples which do not fit the simple ancient-recent trend, and more complex time dependences have been proposed (e.g., Kerridge, 1989). These problems have led to many suggestions that the bulk of the N in the lunar regolith samples is not solar in origin (e.g., Hashizume *et al.*, 2000).

As indicated on Figure 3, the Galileo probe measured $^{15}N/^{14}N$ significantly less for the Jovian atmosphere than the terrestrial atmosphere (e.g., Owen *et al.*, 2001). Given the large uncertainties, the ancient Earth, ancient lunar, and Jupiter data would be consistent with a homogeneous solar system $^{15}N/^{14}N$ roughly 30% lower than the terrestrial atmosphere, but this hypothesis is totally contradicted by the SOHO data. Because of interferences from doubly-charged Si ions, the $^{15}N/^{14}N$ is a difficult measurement for the SOHO instrument. If Genesis data confirm the SOHO, it would suggest a previously-unrecognized major heterogeneity between the materials of the inner and outer solar system.

A large range in $^{15}N/^{14}N$ is observed in meteoritic materials (e.g., Kerridge, 1995). The extreme meteoritic values would go to +150 on the scale of Figure 3. The origins of these variations are totally unknown. Surviving presolar materials produce large variations in meteoritic noble gas isotopic ratios, but the numbers of atoms from pre-solar materials is rather small. Much larger amounts of presolar

N would be required, making the observed variations difficult to understand as pre-solar grain effects. The observed meteoritic $^{15}N/^{14}N$ ratios tend to be higher than the terrestrial atmosphere; this suggests (but certainly does not prove) that meteoritic or cometary inputs might be the source of the high $^{15}N/^{14}N$ measured in recent lunar regolith samples.

In any case the large range of $^{15}N/^{14}N$ shown on Figure 3 represents a serious challenge to the standard model of a homogeneous solar nebula which predicts uniform initial isotopic compositions for solar system objects. An understanding the origins of the isotopic variations shown on Figures 2 and 3 would unquestionably produce greater insight into how planetary materials and objects formed.

5. Analysis Requirements

The levels of analytical sensitivity needed for the various elements and their isotopes depend upon their fluences in the solar wind, on the contamination background expected in the collector materials, and on their relative importance to planetary science. Table I gives our predictions for the fluences of different elements. Analysis of Genesis samples requires sensitivities of parts per million to parts per trillion (depending on the element) in the outer 100 nm of the collector materials surfaces.

The preceding discussion shows the need for data with high precision/accuracy. Table II provides our estimates of the required levels of accuracy and precision required for fundamental advances in understanding planetary materials. Genesis collector materials (see Jurewicz *et al.*, 2003) have been selected and analytical techniques identified which will support reaching the goals set out in Table II.

6. Instrumentation Overview

Figure 4 is a block diagram showing the interdependence of the GENESIS instrumentation and its unique strategy.

In common with all sample return missions, the science objectives are not achieved until the returned samples have been analyzed. This means that not all of the mission instruments are launched. Sample collection instruments are launched (Collector Arrays and Concentrator in our case), but the sample analysis instruments are not. Analytical instruments to be used on the returned samples will certainly include, but not be limited to, noble gas mass spectrometry, static stable isotope mass spectrometry, secondary ion mass spectrometry, resonance ionization mass spectrometry, accelerator mass spectrometry, inductively coupled plasma mass spectrometry, thermal ionization mass spectrometry, and neutron activation analysis.

The spacecraft instruments consist of the following

TABLE I

Estimated composition of bulk solar wind (1)

Z	Element	Solar system abund. (Note 2)	Solar wind flux (cm^{-2} s^{-1})	2-yr Fluence (cm^{-2})	ppma (Note 3)	ppmw (Note 4)
3	Li	5.7E+01	1.7E+00	1.1E+08	2.2E−04	5.3E−05
4	Be	7.3E−01	2.2E−02	1.4E+06	2.8E−06	8.9E−07
5	B	2.1E+01	6.4E−01	4.0E+07	8.0E−05	3.1E−05
6	C	1.0E+07	1.0E+05	6.3E+12	1.3E+01	5.4E+00
7	N	3.1E+06	3.1E+04	2.0E+12	3.9E+00	2.0E+00
8	O	2.4E+07	2.4E+05	1.5E+13	3.0E+01	1.7E+01
9	F	8.4E+02	8.4E+00	5.3E+08	1.1E−03	7.2E−04
10	Ne	3.4E+06	3.4E+04	2.2E+12	4.3E+00	3.1E+00
11	Na	5.7E+04	1.7E+03	1.1E+11	2.2E−01	1.8E−01
12	Mg	1.1E+06	3.2E+04	2.0E+12	4.1E+00	3.5E+00
13	Al	8.5E+04	2.5E+03	1.6E+11	3.2E−01	3.1E−01
14	Si	1.0E+06	3.0E+04	1.9E+12	3.8E+00	3.8E+00
15	P	1.0E+04	2.1E+02	1.3E+10	2.6E−02	2.9E−02
16	S	5.2E+05	1.0E+04	6.5E+11	1.3E+00	1.5E+00
17	Cl	5.2E+03	5.3E+01	3.3E+09	6.7E−03	8.3E−03
18	Ar	1.0E+05	1.0E+03	6.4E+10	1.3E−01	1.7E−01
19	K	3.8E+03	1.1E+02	7.1E+09	1.4E−02	2.0E−02
20	Ca	6.1E+04	1.8E+03	1.2E+11	2.3E−01	3.3E−01
21	Sc	3.4E+01	1.0E+00	6.5E+07	1.3E−04	2.1E−04
22	Ti	2.4E+03	7.2E+01	4.5E+09	9.1E−03	1.5E−02
23	V	2.9E+02	8.8E+00	5.5E+08	1.1E−03	2.0E−03
24	Cr	1.4E+04	4.0E+02	2.6E+10	5.1E−02	9.4E−02
25	Mn	9.6E+03	2.9E+02	1.8E+10	3.6E−02	7.1E−02
26	Fe	9.0E+05	2.7E+04	1.7E+12	3.4E+00	6.8E+00
27	Co	2.2E+03	6.7E+01	4.3E+09	8.5E−03	1.8E−02
28	Ni	4.9E+04	1.5E+03	9.3E+10	1.9E−01	3.9E−01
29	Cu	5.2E+02	1.6E+01	9.9E+08	2.0E−03	4.5E−03
30	Zn	1.3E+03	3.8E+01	2.4E+09	4.8E−03	1.1E−02
31	Ga	3.8E+01	1.1E+00	7.2E+07	1.4E−04	3.5E−04
32	Ge	1.2E+02	3.6E+00	2.3E+08	4.5E−04	1.2E−03
33	As	6.6E+00	2.0E−01	1.2E+07	2.5E−05	6.6E−05
34	Se	6.2E+01	1.9E+00	1.2E+08	2.4E−04	6.6E−04
35	Br	1.2E+01	1.2E−01	7.3E+06	1.5E−05	4.2E−05
36	Kr	4.5E+01	4.5E−01	2.8E+07	5.7E−05	1.7E−04
37	Rb	7.1E+00	2.1E−01	1.3E+07	2.7E−05	8.2E−05

TABLE I
Continued.

Z	Element	Solar system abund. (Note 2)	Solar wind flux (cm^{-2} s^{-1})	2-yr Fluence (cm^{-2})	ppma (Note 3)	ppmw (Note 4)
38	Sr	2.3E+01	7.0E−01	4.4E+07	8.9E−05	2.8E−04
39	Y	4.6E+00	1.4E−01	8.8E+06	1.8E−05	5.6E−05
40	Zr	1.1E+01	3.4E−01	2.2E+07	4.3E−05	1.4E−04
41	Nb	7.0E−01	2.1E−02	1.3E+06	2.6E−06	8.7E−06
42	Mo	2.5E+00	7.6E−02	4.8E+06	9.7E−06	3.3E−05
44	Ru	1.9E+00	5.6E−02	3.5E+06	7.0E−06	2.5E−05
45	Rh	3.4E−01	1.0E−02	6.5E+05	1.3E−06	4.8E−06
46	Pd	1.4E+00	4.2E−02	2.6E+06	5.3E−06	2.0E−05
47	Ag	4.9E−01	1.5E−02	9.2E+05	1.8E−06	7.1E−06
48	Cd	1.6E+00	4.8E−02	3.0E+06	6.1E−06	2.4E−05
49	In	1.8E−01	5.5E−03	3.5E+05	7.0E−07	2.9E−06
50	Sn	3.8E+00	1.1E−01	7.2E+06	1.4E−05	6.1E−05
51	Sb	3.1E−01	9.3E−03	5.8E+05	1.2E−06	5.1E−06
52	Te	4.8E+00	1.4E−01	9.1E+06	1.8E−05	8.3E−05
53	I	9.0E−01	1.8E−02	1.1E+06	2.3E−06	1.0E−05
54	Xe	4.7E+00	4.7E−02	3.0E+06	6.0E−06	2.8E−05
55	Cs	3.7E−01	1.1E−02	6.9E+05	1.4E−06	6.7E−06
56	Ba	4.5E+00	1.3E−01	8.5E+06	1.7E−05	8.3E−05
57	La	4.5E−01	1.3E−02	8.4E+05	1.7E−06	8.3E−06
58	Ce	1.1E+00	3.4E−02	2.2E+06	4.3E−06	2.1E−05
59	Pr	1.7E−01	5.0E−03	3.2E+05	6.3E−07	3.2E−06
60	Nd	8.3E−01	2.5E−02	1.6E+06	3.1E−06	1.6E−05
62	Sm	2.6E−01	7.7E−03	4.9E+05	9.8E−07	5.2E−06
63	Eu	9.7E−02	2.9E−03	1.8E+05	3.7E−07	2.0E−06
64	Gd	3.3E−01	9.9E−03	6.2E+05	1.2E−06	7.0E−06
65	Tb	6.0E−02	1.8E−03	1.1E+05	2.3E−07	1.3E−06
66	Dy	3.9E−01	1.2E−02	7.5E+05	1.5E−06	8.6E−06
67	Ho	8.9E−02	2.7E−03	1.7E+05	3.4E−07	2.0E−06
68	Er	2.5E−01	7.5E−03	4.7E+05	9.5E−07	5.6E−06
69	Tm	3.8E−02	1.1E−03	7.2E+04	1.4E−07	8.6E−07
70	Yb	2.5E−01	7.4E−03	4.7E+05	9.4E−07	5.8E−06
71	Lu	3.7E−02	1.1E−03	6.9E+04	1.4E−07	8.7E−07
72	Hf	1.5E−01	4.6E−03	2.9E+05	5.8E−07	4.2E−06
73	Ta	2.1E−02	6.7E−04	3.9E+04	7.9E−08	9.2E−08
74	W	1.3E−01	4.0E−03	2.5E+05	5.0E−07	3.3E−06
75	Re	5.2E−02	1.6E−03	9.8E+04	2.0E−07	1.3E−06

TABLE I

Continued.

Z	Element	Solar system abund. (Note 2)	Solar wind flux (cm^{-2} s^{-1})	2-yr Fluence (cm^{-2})	ppma (Note 3)	ppmw (Note 4)
76	Os	6.8E−01	2.0E−02	1.3E+06	2.6E−06	1.7E−05
77	Ir	6.6E−01	2.0E−02	1.3E+06	2.5E−06	1.7E−05
78	Pt	1.3E+00	4.0E−02	2.5E+06	5.1E−06	3.5E−05
79	Au	1.9E−01	5.6E−03	3.5E+05	7.1E−07	5.0E−06
80	Hg	3.4E−01	6.7E−03	4.3E+05	8.7E−07	6.1E−06
81	Tl	1.8E−01	5.5E−03	3.5E+05	6.9E−07	5.1E−06
82	Pb	3.2E+00	9.4E−02	6.0E+06	1.2E−05	8.8E−05
83	Bi	1.4E−01	4.3E−03	2.7E+05	5.5E−07	4.0E−06
90	Th	3.4E−02	1.0E−03	6.3E+04	1.3E−07	1.1E−06
92	U	9.0E−03	2.7E−04	1.7E+04	3.4E−08	2.9E−07

Entries are based on a H flux of 3×10^8 $cm^{-2}s^{-1}$ and a solar wind Si/H ratio of 1×10^{-4}. Solar wind abundances for all other elements are calculated by assuming that the relative solar abundances from Anders and Grevesse (1989), except that a correction for first ionization potential depletion (FIP) of 1.5 has been applied for P, S, I, and Hg. A FIP depletion factor of 3 has been applied for C, N, O, F, Ne, Cl, Ar, Br, Kr, and Xe.
Note 1: Entries in this table refer to unconcentrated bulk solar wind.
Note 2: Solar system abundance relative to Si = 10^6.
Note 3: Solar wind concentration averaged over the outer 100 nm of the collector (assumed to be Si) in units of parts per million by number; i.e., (number of solar wind atoms $\times 10^6$)/(atoms of silicon).
Note 4: Solar wind concentration averaged over the outer 100 nm in units of parts per million by weight; i.e., (grams of solar wind element $\times 10^6$)/(grams of silicon).

TABLE II

Precision and accuracy of elemental and isotopic analyses

Elemental accuracy (2σ limits) = $\pm$ 10% of the number of atoms of each element per cm^2 on the collector materials

Isotopic precision (2σ limits on the abundance ratios of the different isotopes of an element compared to a terrestrial reference standard)

C and N	±0.4%
O and Ti	±0.1%
Others	±1%

A special effort will be made to measure the rare gas isotopes, and the abundant ones will be measured to much better than 1%. However, 1% may not be achievable for ^{124}Xe, ^{126}Xe, and ^{78}Kr.

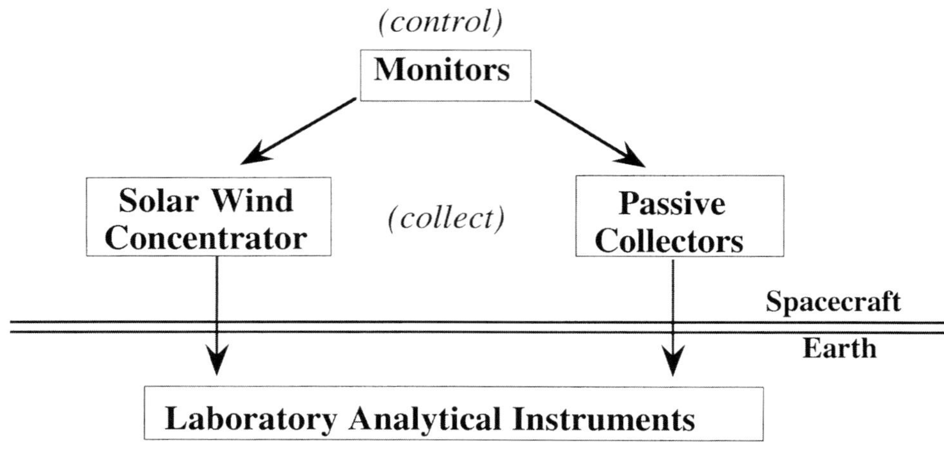

Figure 4. The Genesis instruments are highly integrated towards the focused mission goal of solar wind sample return. The Monitors, electron and ion electrostatic analyzers (Barraclough *et al.*, 2003), yield electron and ion spectra and angular distributions which are used to identify autonomously the prevailing solar wind regime by a Solar Wind Algorithm (Neugebauer *et al.*, 2003). The Algorithm processes the monitor data and then, via the spacecraft Command and Data Handling System, unshades the appropriate specific regime collector array, referred to as 'Passive Collectors' in the figure. The Algorithm also adjusts the voltages on the Concentrator (Nordholt *et al.*, 2003; Wiens *et al.*, 2003) to optimize collection efficiency. Finally, in common with all Sample Return missions, completion of science objectives requires a complement of Laboratory Analytical Instruments which are not launched.

The *Collector Arrays* (passive collectors in Figure 4) are arrays of ultrahigh-purity materials, assembled under clean-room conditions (background for Figure 5), and placed in a clean science *Canister* which is integrated as a unit into the return capsule. Some arrays are always exposed to collect a 'bulk' solar wind sample, while others are deployed only in specific solar wind regimes. Each collector array is made up of 54 full hexagons (approximately 10 cm point to point) and 6 half-hexagons of individual collector materials. More information on the collector array materials is found in the accompanying paper by Jurewicz *et al.*, 2003.

The *Concentrator* is an electrostatic reflecting telescope which mirrors solar-wind ions and concentrates the fluence by a factor of 20 on a set of target materials at the focal point. Ions in the mass/charge range 2.0 to 3.6 amu q^{-1} (He through Mg in terms of elements) are concentrated to obtain high signal-to-background ratios, especially for measurement of N and O isotopes. Such concentration is necessary for measuring these isotopic ratios, which are the highest priority science objectives. The Concentrator is described in detail in the accompanying papers by Nordholt *et al.* (2003), and Wiens *et al.* (2003).

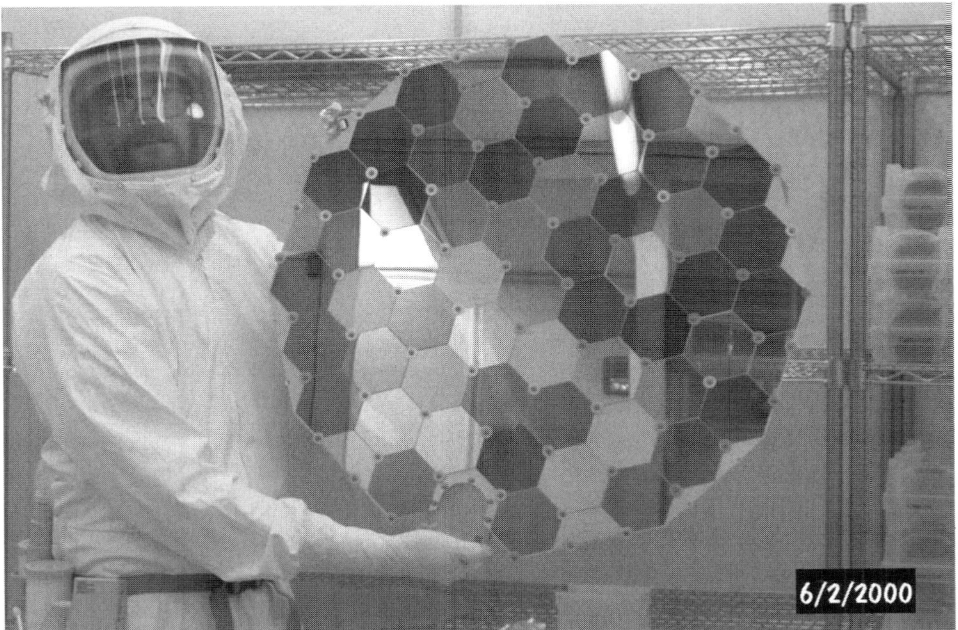

Figure 5. One of the Collector Arrays at the end of integration. This array consists of 54 hexagons, approximately 10 cm point-to-point. The hexagons are made of different collector materials (Jurewicz *et al.*, 2003), optimized for specific measurement objectives. The surfaces are highly polished, resulting in many reflections from objects in the room. To maintain the maximum degree of surface cleanliness for the collector materials, the collector arrays were integrated in a new class 10 clean room at JSC, as can be seen in the background of this figure as well as Figures 4 and 5.

Recognition of different solar wind regimes is accomplished with standard solar wind ion and electron *Monitors* (Barraclough *et al.*, 2003). The output signals of the monitors are analyzed by a 'science algorithm' (Neugebauer *et al.*, 2003) to determine the prevailing solar wind regime and to deploy autonomously the appropriate Collector Array. The output of the monitors is also used to set voltage levels in the Concentrator and to determine the total fluence of H and He to the arrays.

The Collector Arrays are contained in the *Canister* (Figures 4 and 5). As shown on Figure 6, there are a total of 5 arrays, one of which is fixed in the Canister Cover and the other four deployable arrays vertically stacked. The 4 deployable arrays are rotated clockwise 256 deg to expose the Concentrator to the solar wind. The fixed array and the top array of the stack are always exposed and thus collect a bulk solar wind sample. Depending on the prevailing solar wind regime, as determined by the monitors, one of the three bottom arrays in the stack is unshaded by an 152 deg counterclockwise rotation on the shaft of the Array Deployment Mechanism relative to the configuration shown in Figures 4 or 5. When the monitors record a change in the solar wind regime, the previously unshaded array is shielded from

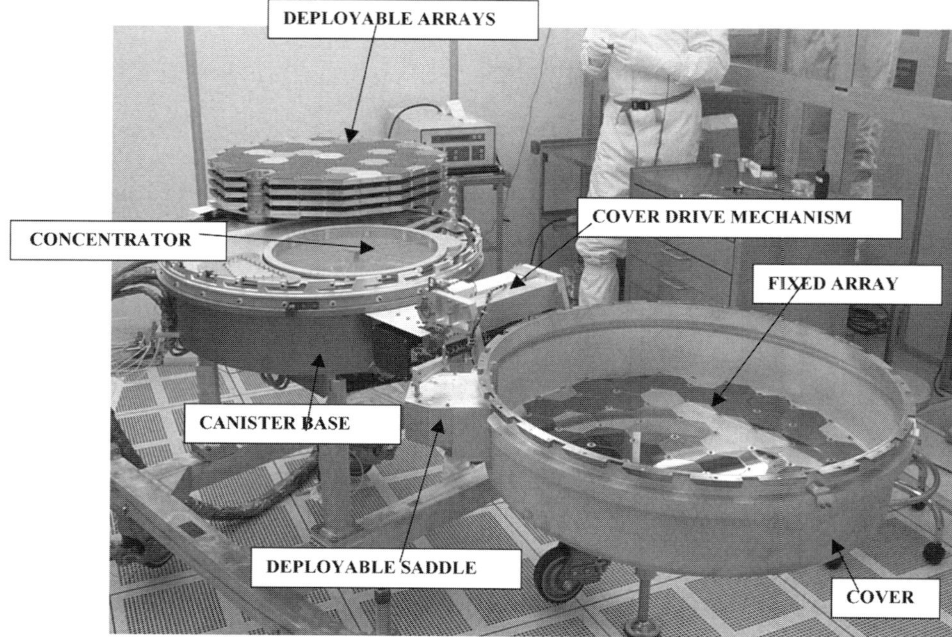

Figure 6. Under class 10 clean room conditions, the Collector Arrays and Concentrator are integrated into a Canister. The Canister preserves the cleanliness of the collector materials. After functional testing, the Canister Cover was closed in the clean room and not re-opened until ready to collect solar wind at L1. The locations of the Fixed Array in the Canister Cover and the Deployable Array stack show clearly in this image. In flight, one of the three lower arrays in the stack is 'unshaded', to provide an independent sample of a given solar wind regime. The selection of the regime and deployment of the arrays is done autonomously, based on data from the Monitors.

the solar wind by being placed back in the stack and the appropriate new array unshaded.

At the completion of solar wind collection in April 2004, the stack of deployable arrays will be stowed by rotation back inside the canister. The Cover Drive Mechanism will be activated to close the Canister Cover. The Lock Ring Drive Mechanism is rotated to seal the Cover to the Canister Base and to latch it for the return back to Earth. Inside the canister, each Array is firmly held between the Fixed Saddle (Figure 7) and the Deployable Saddle (Figure 6), which is rotated inward against the Array Stack by the action of the Lock Ring Drive to clamp the Arrays.

To prevent contamination on Earth return, a Filter (not shown) is installed in the Canister Base. This has been tested to demonstrate removal of particulates greater than 0.3 microns in size and to effectively getter gases. The Canister was placed on a high purity nitrogen gas purge from the end of integration until a few hours before launch. This purge will be re-established as soon as possible after recovery.

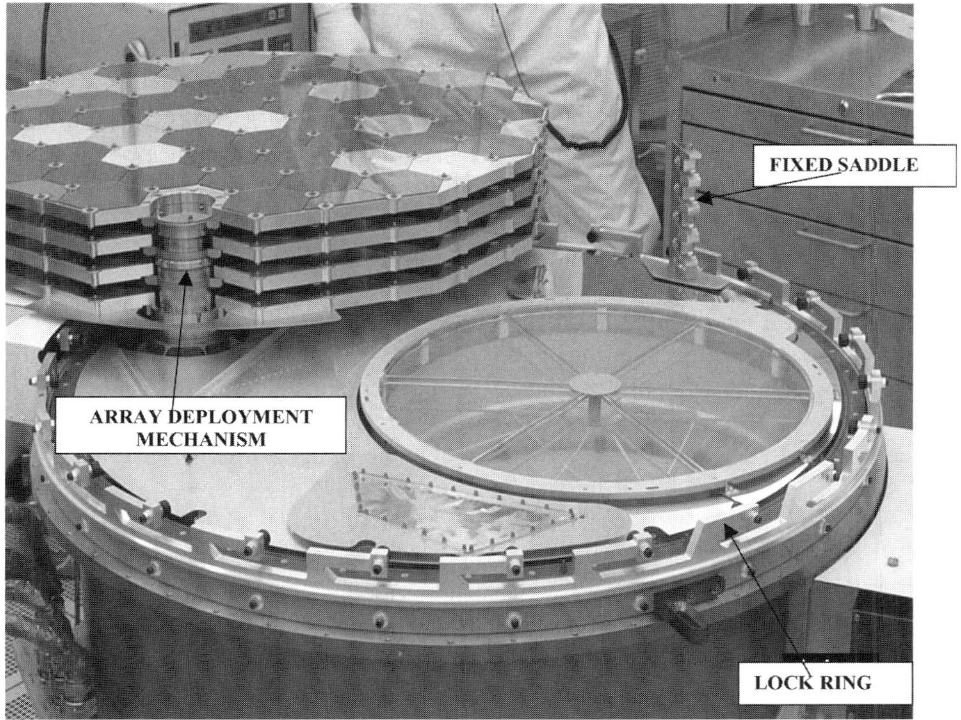

Figure 7. A closer view of the concentrator and Deployable Array stack in the deployed configuration. To sample a given solar wind regime, one of the three lower arrays in the stack is unshaded by rotating back in the direction of the Concentrator. The rotation is accomplished by motors (not shown) activating the Array Deployment Mechanism. When the Deployable Arrays are stowed and the Canister Cover closed, the array stack is held securely at two additional points besides the Array Deployment Mechanism shaft: (1) the Fixed Saddle fastened to the canister base as shown here and (2) the Deployable Saddle in the Canister Cover, as shown in Figure 6. After the cover is closed, the Deployable Saddle rotates out from its housing in the cover to clamp the arrays. Finally the Lock Ring is activated to seal the Canister Cover and Base. When the arrays are deployed, two additional areas in the canister are exposed to the solar wind. A Au foil collector can be seen in the lower center of the figure. A polished Al collector can be seen between the Concentrator and the Fixed Saddle.

The strict cleanliness requirements levied on the Canister and its contents demanded changes from the normal development and test flow for spacecraft hardware. The flight model Canister was built and tested using non-flight collectors identical to those being flown. After complete testing, including vibration and solar-thermal vacuum testing, the Canister was completely disassembled. The structure and array frames were cleaned in a class-10 cleanroom, re-assembled, and the array frames were populated with fresh, clean collectors. After installation of the concentrator, the Canister was closed in the class-10 cleanroom and was never opened again until in space.

7. Spacecraft

The Genesis spacecraft was designed to be a sun-pointing major axis spinner, consisting of an equipment deck and a sample return capsule (SRC). The equipment deck provides structural support for the spacecraft subsystems, the Ion and Electron monitors and the SRC. The SRC contains the science canister and its payload described above.

Figure 8 shows the overall spacecraft with the SRC attached to the upper side of the deck. The two solar arrays are capable of generating 281 W at 1.012 AU, 10° off-sun, 65 °C at end of life. A rechargeable 16 amp hr^{-1}, 28 V spacecraft battery provides power during maneuvers if the spacecraft is pointed away from the sun. Redundant low gain antennas are mounted to the solar arrays on both front and aft sides. Redundant transponders are used to support communications with the Deep Space Network (DSN) stations. The two propellant tanks visible at the edge of the deck each hold more than 71 kg of useable hydrazine. Figure 8 also highlights the lower deck, or underside, of the spacecraft. The spacecraft battery is located inside the launch vehicle adapter ring. The medium gain antenna provides high data rate communications with the DSN during science data transmission. The spacecraft employs passive thermal control plus autonomous and ground-controlled heaters.

The lower portion of Figure 8 also shows various elements of the attitude control subsystem (ACS). To avoid contamination on the collector arrays, all thrusters are below the spacecraft deck, out of any line of sight of the SRC. Eight small (0.9 N) reaction control system thrusters, canted from the spin axis, are used to control the spin rate, precess the spin axis daily, and perform small delta-V maneuvers. Large delta V maneuvers are performed with four large (22 N) trajectory correction maneuver thrusters, axially directed.

There are three attitude sensor types: digital 2-axis sun sensor (DSS), spinning Sun sensor (SSS) and star tracker (ST). The spinning Sun sensor measures the sun crossing angle. The digital sun sensor is a two-axis version of the SSS, but mounted with a view along the spin axis. SSS and DSS processed output provides spin axis off-sun angle and spin rate. Under high nutation and/or sun proximity to the spin axis, multiple sun crossings can occur at irregular time intervals. These can severely degrade knowledge of the spin rate. Sun pointing keep out zones were established to avoid this occurrence. Mission flight rules, and both flight and ground software, enforce the keep out zones.

ACS is also responsible for the passive nutation damper system. This system consists of a viscous fluid filled tube that surrounds each propulsion tank and fluid control system. The shorter the nutation time constant, the quicker nutation is damped out. In general, the time constants diminish as fuel is expended and/or spin rate is increased. A typical time constant is 2.5 hours when the spacecraft has half its fuel left, in the science configuration, at the nominal 1.6 rpm spin rate.

A unique challenge for Genesis was to design the spacecraft so its principal axis remained aligned with the spin axis in several different configurations, minimizing

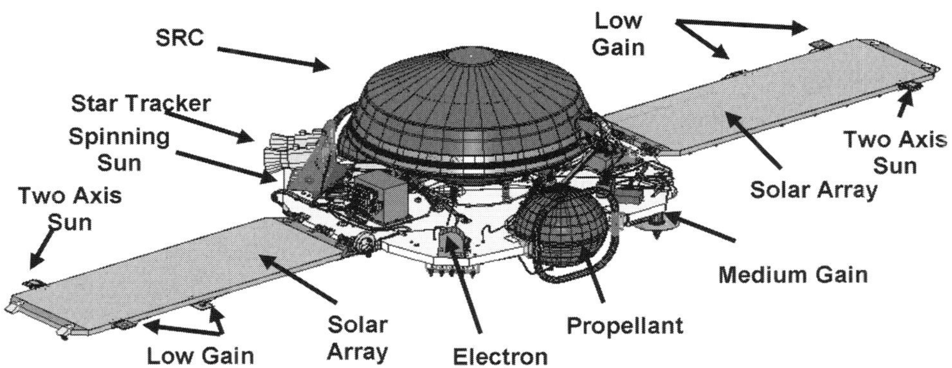

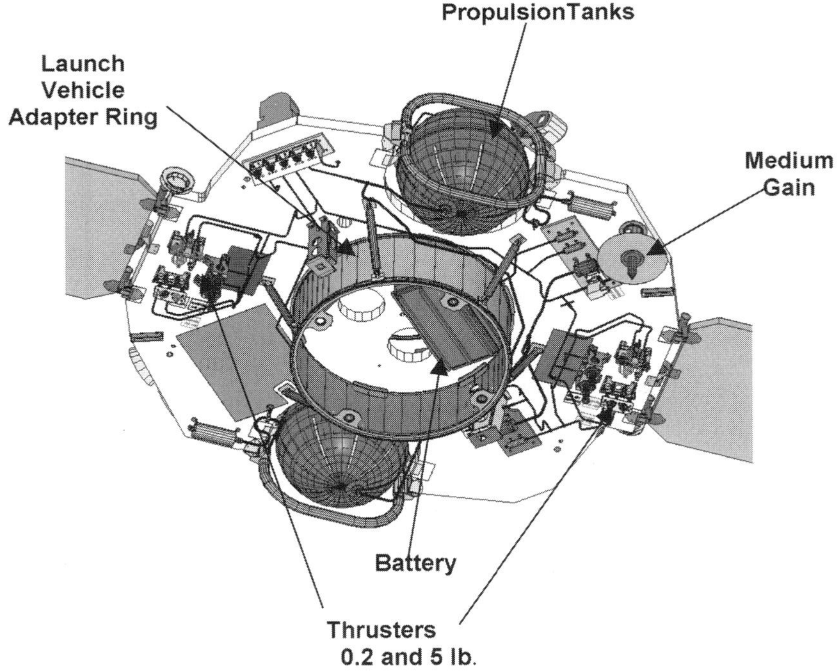

Figure 8. Two views of the Genesis spacecraft, as seen from the Sun (upper) and from the Earth (lower). The overall flight system consists of the Sample Return Capsule (SRC) which houses the Canister and Collector Materials (Figures 3–5), a flat spacecraft deck on which the Monitors and most of the engineering subsystems are mounted, and the solar arrays. The SRC is shown in the closed configuration present just after launch or just prior to re-entry. While at L1, the Sun-facing backshell opens to expose the Canister.

wobble. The initial and final configurations are with the SRC closed. Alignment in this configuration is driven by the need for accurate pointing when the SRC is released for re-entry. The post-launch check-out configuration had the SRC lid open but the science canister closed prior to L1 orbit insertion. And finally, in the science collection phase the canister lid is open and the arrays are deployed. The position of the arrays also affect spacecraft wobble. Adjustment of the SRC lid angle aids in minimizing wobble among the various configurations in which the lid is open. The SRC lid angle is currently 192.2° in the science-collection phase, and was 192.9° during portions of the check-out phase. Analysis of the first several months of operation shows that the wobble remains under 0.35° in the science-collection orientation, and under 0.2° in all other orientations. This is sufficient to meet the (wobble+nutation) requirement, which is driven by the ion temperature measurement.

The Command and Data Handling (C&DH) subsystem is housed on the spacecraft's forward deck. The C&DH provides time definition and command and data interfaces with all other subsystems. It is fully redundant and single fault tolerant. In operation, it has proved quite resistant to solar proton events. The C&DH contains multiple processor and memory cards and virtual memory buses. Some of the principal cards are the Flight Processors that contain the central processor unit, dynamic random access memory (DRAM), two different backup memory devices, the Payload and Pointing Interface Card (PPIC), and the Command Module Interface Card (CMIC). Flight Software (FSW) includes the onboard code which runs the spacecraft, including fault protection. FSW has a selectable operating speed and utilizes less than 60% of processor capability. 128 MB (megabytes, $1024 \times 1024 \times 8$ bits) of DRAM is used for all operations. FSW has been allocated 32 MB, and telemetry storage has been allocated 96 MB for science and engineering data. This is sufficient for multiple playbacks in the current downlink strategy.

Fault protection is responsible for failure detection, response and recovery. A hierarchical detection strategy isolates the failure. Responses include switching from primary to redundant strings and/or swapping C&DH sides. If needed, an autonomous safe mode is entered which reconfigures the vehicle to minimize electrical power loads, continues fault detection and response, and (if the spacecraft is $> 35°$ off sun) precesses quickly to a sun pointed attitude for solar array power.

The original mission plan called for spacecraft contact with the DSN only once per week during the science phase of the mission. A more conservative strategy was eventually implemented in which contact is established approximately three times per week during routine portions of the mission and more often during high-activity phases. This allows multiple playback opportunities for the data and more frequent monitoring of the spacecraft's state of health. Upon transmission to Earth, the data are received by either 26-m or 34-m DSN antennas. There are 8 downlink data rates, from 1050 to 47400 bits per second, depending on spacecraft antenna, DSN antenna, and mission phase.

The sample return capsule (SRC) contains all of the payload to be returned to Earth, as well as re-entry support systems. It is a clamshell design, built to allow exposure of the collector materials in space and protect them during re-entry. The upper half of the clamshell contains the descent parachute. The lower half, enclosed by the heat shield, contains the payload canister surrounded by the avionics components required for recovery of the capsule. These include a patch antenna, VHF locator beacon, GPS receiver, UHF transceiver, SRC battery, and supporting electronics. The GPS receiver is used to give the position of the incoming capsule once the parachute is deployed to support mid-air capture by helicopter. The locator beacon is included as a backup, in the event of a contingency ground recovery. The beacon signal is intended to last a number of hours after landing. Details on capsule recovery are given below.

More details on spacecraft subsystems are given in Hong *et al.* (2002).

8. Mission Operations and Recovery

The GENESIS mission trajectory is shown in Figure 1. Following launch on August 8, 2001, the GENESIS spacecraft underwent a check-out period during approach to the L1 position. The monitors were turned on August 23, 2001 and in-situ testing of the science algorithm commenced. A check-out phase of approximately three months prior to opening the Canister allowed several solar rotations over which to test the identification of the different solar-wind regimes (Neugebauer *et al.*, 2003). This time was also used to outgas the spacecraft in order to minimize any possible contamination to the collector arrays. The return capsule lid was open either partially or fully during this time to facilitate outgassing.

The spacecraft underwent L1 orbit insertion on November 19, 2001, and the Canister was opened shortly thereafter, on November 30, to begin solar-wind collection. During the collection phase, the spacecraft maintains an orientation with the spin axis pointed 4.5 ± 2.0 deg ahead of the sun, which is the average apparent direction of the solar wind, considering spacecraft motion around the Sun. The spacecraft autonomously performs a one-degree precession maneuver each day to maintain this position as it orbits the Sun along with the Earth. The spacecraft also maintains a spin rate of 1.6 rpm $\pm$ 10%, which allows the monitors to sweep over 360° in the azimuth angle about the spin axis at regular intervals. The collection phase is slated to continue until April, 2004 with only minor interruptions for station-keeping maneuvers. These maneuvers occur approximately six times per year, expending only 1–2 days each of potential collection time. Thrusters used for these maneuvers and also for daily precession maneuvers are all biased away from the front of the spacecraft – and from the collector arrays – to minimize potential contamination to the arrays during thruster firings.

At the end of the collection period in April 2004, the Canister and SRC are closed, and a small maneuver begins the journey back to Earth. Figure 1 shows that instead of returning directly to Earth, the spacecraft takes a more circuitous route through the L2 region before re-entry in order to allow re-entry to occur at relatively high latitude at the Utah Test and Training Range and on the daylit side of the Earth. In case of bad weather in Utah, (which would be unusual for this time of year) diversion to a 'parking orbit' is possible which can delay the spacecraft's reentry for up to 19 days.

Before reentry, the spacecraft will assume a nose-down reentry attitude. The capsule stabilizes with its nose down because of the location of its center of gravity, its spin rate and aerodynamic shape. The spacecraft spins up to about 16 revolutions per minute, then the SRC separates. The remainder of the spacecraft will execute a deboost maneuver so that it reenters Earth's atmosphere and burns up over the Pacific Ocean. The SRC will enter Earth's atmosphere at an entry angle of minus 8 deg at Earth escape velocity (11.04 km per second).

At 30 km altitude, the capsule will deploy a drogue parachute to begin slowing its descent. That in turn will pull out the main parachute, a parafoil which descends in a broad spiral producing a low vertical descent velocity. The parafoil size determines its airspeed, which is chosen for best mid-air retrieval safety and reliability.

Mid-air retrieval is accomplished by helicopter snatch. Two helicopters (primary and backup) will provide up to 5 passes. Radar and infra-red tracking will guide the helicopters to the SRC. The mid-air retrieval subsystem consists of a constant tension winch, a single pole that pivots off the pilot-side landing skid, a hook-and-release mechanism and rigging to guide the retrieval cable out the door of the helicopter. Practice trials have always been successful on the first pass.

The landing site at the Utah Test and Training Range was chosen because the area is a vast, unoccupied salt flat controlled by the U.S. Army and Air Force. The landing footprint for the sample return capsule will be about 30 by 84 km (18 by 52 miles), an ample area to allow for aerodynamic uncertainties and winds that might affect the direction the capsule travels in the atmosphere. The sample return capsule will approach the landing zone on a heading of approximately 122 deg on a northwest to southeast trajectory. Landing is planned to take place at 9 AM local time.

9. Sample Allocation and Curation

Following recovery, the canister containing the collector arrays and the concentrator will be taken to the receiving and curatorial facility at Johnson Space Center (JSC), the designated NASA Center for curation of extraterrestrial materials.

The Canister was cleaned and the Collector Arrays and Concentrator integrated in a new Class 10 cleanroom at JSC (background for Figure 5) built specifically for the Genesis mission. Post-recovery operations will be carried out in this same facility. Inasmuch as possible, the contamination control procedures during canister disassembly will be identical to those used in canister assembly. In addition, systematic, archeological-style inspection of the returned materials and components will be carried out to determine what actually happened during exposure. Based on visual inspection, any new surface marks, specifically locations of micrometeorite impacts, will be documented.

Upon receipt of the canister at JSC the transport container will be opened in a class 10 000 area. The exterior of the science canister will be wiped down with solvent until visibly clean prior to introducing the canister into the class 100 sample extraction area. Additional cleaning techniques may be used depending on the condition of the canister. The science canister will be opened in the class 100 facility and the condition of the canister interior will be described and imaged.

The disassembled collectors and concentrator target materials will be handled in air in the class 10 cleanroom conditions and stored in a dust-free, contaminant-free, high purity nitrogen environment. The collector and target surfaces will be protected from all physical contact and from static electric charging. For the specific case of Si wafers, well defined handling procedures developed by the semiconductor industry have been adapted to our needs. Procedures for subdividing individual collectors for allocation without significant contamination are being devised and tested by JSC in consultation with the Science Team.

Before allocation, the samples will be examined for factors that might affect end-use analyses, and results of those examinations will be a part of the documentation associated with each sample. The locations of the samples and the investigators to whom they are allocated will be known at all times. These data will be readily available and secure.

10. Overview of Plans for Sample Analysis

The opportunity to analyze returned solar wind samples will be open to the international planetary materials community as soon as possible after recovery. Allocations will be made after careful review of the proposed analytical procedures by a mission-independent Sample Allocation Committee (SAC). It is also desirable to have a focused and timely product of the mission. As we are confident that several of the important science objectives can be realized relatively quickly without compromising science quality, four studies – N isotopes, noble gas isotopes in bulk solar wind, C isotopes, and a search for radioactive nuclei – will be set aside to be performed by the Genesis Co-Investigators as an Early Science Return. These Early Science Return measurements were selected on the basis of a combination of science importance and feasibility. Less than 1% of the returned sample from the canister will be used in these studies.

Materials for study will be provided by the Curatorial Facility of the Johnson Space Center based on recommendations of the SAC. This process follows well-developed procedures with deep heritage going back to Apollo lunar sample allocations. In addition to SAC operation, these procedures cover the selection of SAC members, the frequency of meetings, etc. The SAC also serves as a monitoring and advisory committee to the JSC Curatorial Facility on issues relating to minimizing contamination during the handling and storage of collector materials.

Acknowledgements

The authors of this overview obviously only represent a small fraction of the skillful and dedicated technical staff from the Jet Propulsion Laboratory, Lockheed Martin Astronautics, Johnson Space Center, and Los Alamos National Laboratory who co-operated to make the Genesis mission a success. We gratefully acknowledge the support of this team. We also acknowledge the support and oversight provided by upper management at the Jet Propulsion Laboratory, Lockheed Martin Astronautics and NASA Headquarters.

References

Anders, E., and Grevesse, N.: 1989, 'Abundances of the Elements: Meteoritic and solar', *Geochim. Cosmochim. Acta* **53**, 197–214.

Barraclough, B. L., Dors, E. E., Abeyta, R. A., Alexander J. F., Ameduri, F. P. Baldonado, J. R. Bame, S. J., Casey, P. J., Dirks, G., Everett, D. T., Gosling, J. T., Grace, K. M., Guerrero, D. R., Kolar, J. D., Kroesche, J., Lockhart, W., McComas, D. J., Mietz D. E., Roese, J., Sanders, J., Steinberg, J. T., Tokar, R. L., Urdiales, C., and Wiens, R. C.: 2003, 'The Plasma Ion and Electron Instruments for the Genesis Mission', *Space Sci. Rev.*, this volume.

Bochsler, P.; 2000, 'Abundances and Charge States of Particles in the Solar Wind.' *Rev. Geophys.* **38**, 247–266.

Burnett, D. S., Woolum, D. S., Benjamin, T. M., Rogers, P. S. Z., Duffy, C. J., and Maggiore, C.: 1989, 'A Test of the Smoothness of the Elemental Abundances of Carbonaceous Chondrites.' *Geochim. Cosmochim. Acta* **53**, 471–481.

Clayton, R. N.: 1993, 'Oxygen Isotopes in Meteorites', *Ann. Rev. Earth Planetary Sci.* **31**, 115–149.

Clayton, R. N.: 2002, 'Self-Shielding in the Solar Nebula', *Nature* **415**, 860–861.

Clayton, R. N. and Mayeda, T. K.: 1984, 'The O Isotope Record in Murchison and Other Carbonaceous Chondrites', *Earth Planetary Sci. Lett.* **67**, 151–161.

Collier, M. R., Hamilton, D. C., Gloeckler, G., Ho, G., Bochsler, P., Bodmer, R., and Sheldon, R.: 1998, 'Oxygen 16 to Oxygen 18 Abundance Ratio in the Solar Wind Observed by Wind/MASS', *J. Geophys. Res.* **103**, 7–13.

Fisk, L. A., Schwadron, N. A., and Zurbuchen, T. H.: 1998, 'On the Slow Solar Wind', *Space Sci. Rev.* **86**, 51–60.

Geiss, J., Buehler F., Cerutti H., Eberhardt P., and Filleux C.: 1972, 'Solar Wind Composition Experiment,' *Apollo 16 Preliminary Science Report*, NASA SP-315, pp. 14-1–14-10.

Hashizume, K., Chaussidon, M., Marty, B., and Robert, F.: 2000, 'Solar Wind Record on the Moon: Deciphering Presolar from Planetary Nitrogen', *Science* **290**, 1142–1145.

Hong, P. E., Carlisle, G., and Smith N. G.: 2002, 'Look, Ma, No HANS', *Proceedings of the 2002 IEEE Aerospace Conference*, to appear.

Jurewicz, A. J. G.,.Burnett, D. S., Wiens, R. C., Friedmann, T. A., Hays, C. C., Hohlfelder, R. J., Nishiizumi, K., Stone, J. A., Woolum, D. S., Becker, R., Butterworth, A. L., Campbell, A. J.,

Ebihara, M., Franchi, I. A., Heber, V., Hohenberg, C. M., Humayun, M., McKeegan, K. D., McNamara, K., Meshik, A., Pepin, R. O., Schlutter, D., and Wieler, R.: 2003, 'Overview of the Genesis Solar-Wind Collector Materials', *Space Sci. Rev.*, this volume.

Kallenbach, R.: 2001, 'Isotopic Composition Measured *in-situ* in Different Solar Wind Regimes by CELIAS/MTOF on SOHO', in R. F. Wimmer-Schweingruber (ed.), *Solar and Galactic Composition* Am. Inst. of Physics, pp. 113–119.

Kallenbach, R., Ipavich, F. M., Kucharek, H., Bochsler, P., Galvin, A. B., Geiss, J., Gliem, F., Gloeckler, G., Grünwaldt, H., Hefti, S., Hilchenbach, M., and Hovestadt, D.: 1998a, 'Fractionation of SI, NE and MG Isotopes in the Solar Wind as Measured by SOHO/CELIAS/MTOF', *Space Sci. Rev.* **85**, 357–370.

Kallenbach, R., Geiss, J., Ipavich, F. M., Gloeckler, G., Bochsler, P., Gliem, F., Hefti, S., Hilchenbach, M., and Hovestadt, D.: 1998b, 'Isotopic Composition of Solar Wind Nitrogen: First *in situ* Determination with the CELIAS/MTOF Spectrometer on Board SOHO', *Astrophys. J.* **507**, L185 –L188.

Kallenbach, R., Ipavich, F. M., Kucharek, H., Bochsler, P., Galvin, A. B., Geiss, J., Gliem, F., Gloeckler, G., Grünwaldt, H., Hilchenbach, M., and Hovestadt, D.: 1999, 'Solar Wind Isotopic Abundance Ratios of Ne, Mg and Si Measured by SOHO/CELIAS/MTOF as Diagnostic Tool for the Inner Corona', *Phys. Chem. Earth (C)* **24**, 415–419.

Kerridge, J. F.: 1989, 'What Has Caused the Secular Increase in ^{15}N?', *Science* **245**, 480–486.

Kerridge, J. F.: 1995, 'Nitrogen and its Isotopes in the Early Solar System', in K. A. Farley (ed.), *Volatiles in the Earth and Solar System Conf. Proc.* **341**, Am. Inst. of Physics, pp. 167–174.

Marsch, E., von Steiger, R., and Bochsler, P.: 1995, 'Element Fractionation by Diffusion in the Solar Chromosphere', *Astron. Astrophys.* **301**, 261–276.

McSween, H. Y.: 1993, 'Cosmic or Cosmuck?', *Meteoritics* **28**, 3–4.

Neugebauer, M.: 1991, 'The Quasi-Stationery and Transient States of the Solar Wind', *Science* **252**, 404–409.

Neugebauer, M., Steinberg, J. T., Tokar, R. L., Barraclough, B. L., Dors, E. E., Wiens, R. C., Gingerich, D. E., Luckey, D., and Whiteaker, D. B.: 2003, 'Genesis On-Board Determination of the Solar Wind Flow Regime', *Space Sci. Rev.*, this volume.

Nordholt, J. E., Wiens, R. C., Abeyta, R. A., Baldonado, J. R., Burnett, D. S., Casey, P., Everett, D. T., Kroesche, J., Lockhart, W., McComas, D. J., Mietz, D. E., MacNeal, P., Mireles, V., Moses, R. W. Jr., Neugebauer, M., Poths, J., Reisenfeld, D. B., Storms, S. A., and Urdiales, C.: 2003, 'The Genesis Solar Wind Concentrator', *Space Sci. Rev.*, this volume.

Owen, T., Mahaffy, P. R., Niemann, H. B., Atreya, S., and Wong, M: 2001, 'Protosolar Nitrogen', *Astrophys. J. Lett.* **553**, L77–L79.

Ozima, M., Wieler, R., Marty, B., and Podosek, F. A.: 1998, 'Comparative Studies of Solar, Q-Gases and Terrestrial Noble Gases, and Implications on the Evolution of the Solar Nebula', *Geochim. Cosmochim. Acata* **62**, 301–314.

Pepin, R. O.: 1991, 'On the Origin and Evolution of Terrestrial Planet Atmospheres and Meteoritic Volatilies', *Icarus* **92**, 2–79.

von Steiger, R., Schwadron, N. A., Fisk, L. A., Geiss, J., Gloeckler, G., Hefti, S., Wilken, B., Wimmer-Schweingruber, R. F., and Zurbuchen, T. H.: 2000, 'Composition of Quasi-Stationary Solar Wind Flows from Ulysses/Solar Wind Ion Composition Spectrometer', *J. Geophys. Res.* **105**, 27217–27238.

Wieler, R., Kehm, K., Meshik., A. P., and Hohenberg, C. M.: 1996, 'Secular Changes in the Xenon a nd Krypton Abundances in the Solar Wind Recorded in Single Lunar Grains', *Nature* **384**, 46–49.

Wiens, R. C., Huss, G. R., and Burnett, D. S.: 1999, 'The Solar Oxygen-Isotopic Composition: Predictions and Implications for Solar Nebula Processes', *Met. Planetary Sci.* **34**, 99–107.

Wiens, R. C., Neugebauer, M., Reisenfeld, D. B., Moses, R. W. Jr., and Nordholt, J. E.: 2003, 'Genesis Solar Wind Concentrator: Computer Simulations of Performance under Solar Wind Conditions', *Space Sci. Rev.*, this volume.

Wimmer-Schweingruber, R. F. and Bochsler, P.: 2001, 'The Isotopic Composition of Oxygen in the Fast Solar Wind: ACE/SWIMS', *Geophys. Res. Lett.* **28**, 2763–2766.

THE GENESIS SOLAR-WIND COLLECTOR MATERIALS

A. J. G. JUREWICZ[1], D. S. BURNETT[2], R. C. WIENS[3], T. A. FRIEDMANN[4], C. C. HAYS[1,2], R. J. HOHLFELDER[4], K. NISHIIZUMI[5], J. A. STONE[1], D. S. WOOLUM[6], R. BECKER[7], A. L. BUTTERWORTH[8], A. J. CAMPBELL[9], M. EBIHARA[10], I. A. FRANCHI[8], V. HEBER[11], C. M. HOHENBERG[12], M. HUMAYUN[9], K. D. MCKEEGAN[13], K. MCNAMARA[14], A. MESHIK[12], R. O. PEPIN[7], D. SCHLUTTER[7] and R. WIELER[11]

[1]*Jet Propulsion Laboratory/California Institute of Technology, m/s 183-501, 4800 Oak Grove Dr., Pasadena CA 91109-8099, U.S.A.*
[2]*California Institute of Technology, Pasadena CA, U.S.A.*
[3]*Los Alamos National Laboratory, Los Alamos, NM, U.S.A.*
[4]*Sandia National Labs. Albuquerque, NM, U.S.A.*
[5]*Space Sciences Lab. U. Calif, Berkeley, CA, U.S.A.*
[6]*Dept. of Physics, Calif. St. U, Fullerton, CA, U.S.A.*
[7]*Dept. of Physics, U. Minnesota, Minneapolis, MN, U.S.A.*
[8]*Planetary Science Research Institute, Open U, Milton Keynes, U.K.*
[9]*Dept. of Geophysical Sciences, U. of Chicago, Chicago, IL, U.S.A.*
[10]*Dept. of Chemistry, Tokyo Metropolitan University, U.S.A.*
[11]*Institute for Isotope Geology and Mineral Resources, ETH Zürich, Switzerland*
[12]*Dept. Of Physics, Washington U. St. Louis, MO, U.S.A.*
[13]*Dept. of Earth and Space Sciences, UCLA, Los Angeles, CA, U.S.A.*
[14]*NASA Johnson Space Center, Houston, TX, U.S.A.*
(Author for correspondence, E-mail: Amy.J.Jurewicz@jpl.nasa.gov)

Received 30 January 2002; Accepted in final form 21 May 2002

Abstract. Genesis (NASA Discovery Mission #5) is a sample return mission. Collectors comprised of ultra-high purity materials will be exposed to the solar wind and then returned to Earth for laboratory analysis. There is a suite of fifteen types of ultra-pure materials distributed among several locations. Most of the materials are mounted on deployable panels ('collector arrays'), with some as targets in the focal spot of an electrostatic mirror (the 'concentrator'). Other materials are strategically placed on the spacecraft as additional targets of opportunity to maximize the area for solar-wind collection.

Most of the collection area consists of hexagonal collectors in the arrays; approximately half are silicon, the rest are for solar-wind components not retained and/or not easily measured in silicon. There are a variety of materials both in collector arrays and elsewhere targeted for the analyses of specific solar-wind components.

Engineering and science factors drove the selection process. Engineering required testing of physical properties such as the ability to withstand shaking on launch and thermal cycling during deployment. Science constraints included bulk purity, surface and interface cleanliness, retentiveness with respect to individual solar-wind components, and availability.

A detailed report of material parameters planned as a resource for choosing materials for study will be published on a Genesis website, and will be updated as additional information is obtained. Some material is already linked to the Genesis plasma data website (genesis.lanl.gov). Genesis should provide a reservoir of materials for allocation to the scientific community throughout the 21st Century.

Space Science Reviews **105:** 535–560, 2003.

1. Introduction

The concept behind Genesis, the solar-wind sample-return Discovery Mission, is simply to expose materials to the solar wind and to bring them back to Earth for laboratory analysis. Accordingly, the heart of Genesis is the suite of ultra-pure materials used as collectors.

Fifteen high-purity materials were chosen using criteria that will eventually allow for high-precision laboratory analyses of specific solar-wind elements or isotopes in each collector. The most important selection criteria were: bulk purity, surface cleanliness, and the retentiveness of solar-wind components under mission conditions. Further, these flight-qualified materials can physically withstand the environment the spacecraft sees during launch, cruise, and landing phases. Finally, there was an effort to insure that the collector-material designated for a given element (or isotope) was compatible with the technique(s) most likely to be used for the eventual, high-precision laboratory analysis.

Conveniently, the semiconductor industry has pioneered manufacturing and handling techniques for many ultra-pure materials. In fact, most of the materials used in Genesis were commercially available (eg., silicon, germanium, and sapphire). A few collector materials were not commercially available but have commercial counterparts. These flight-materials were especially tailored for the Genesis mission (e.g., ^{13}C-diamond, diamond-like carbon, bulk metallic glass). The rest were thin-film coatings produced under the auspices of the Genesis science team. Control samples of the flight materials will be curated at the Johnson Space Center (JSC) and allocated for testing purposes. Similarly, all relevant portions of the spacecraft will be archived for additional solar wind, micrometeorite, and other future studies.

The Genesis collector materials are deployed in several locations: the concentrator; the five collector arrays; the two side collectors; the top of the array-deployment mechanism, and the sample-return capsule lid-foil collector. All materials are passive collectors in the sense that when the materials are exposed to the solar wind, the ions will implant themselves into the materials. However, in the concentrator, solar wind is reduced in hydrogen and helium, and an applied voltage concentrates the solar-wind ions in the 4 to 28 amu range before they implant themselves into the collector materials. The concentrator increases the fluence on the target collector materials by a factor of 20, thus facilitating the isotopic analysis of light elements (Nordholt *et al.*, 2003; Wiens *et al.*, 2003). The locations where collector materials are deployed are given in Figures 1–3. Note that, except for the lid-foils, all of the collectors are contained in a cylindrical 'canister'. The canister fits inside a sample-return capsule, the exposed portion of which is lined by lid-foil collectors. Both the canister and the sample-return capsule must be open for solar-wind collection (Burnett *et al.*, 2003 and Figure 3). Most materials will be exposed to the solar wind throughout the mission, and so will bring back bulk solar-wind samples. However, three collector arrays will be deployed (i.e., exposed

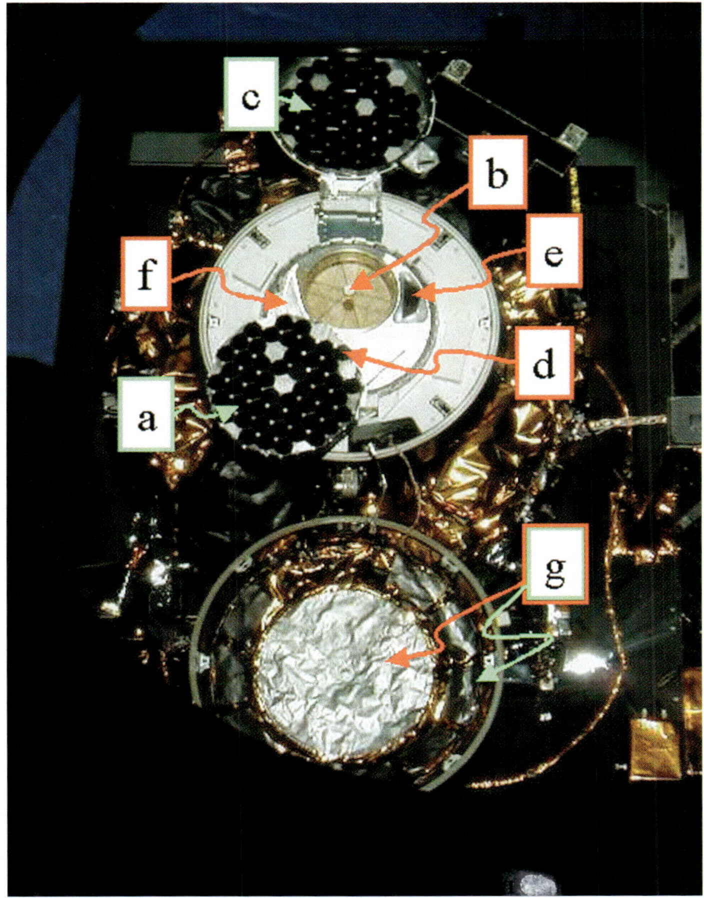

Figure 1. Genesis sample-return capsule during thermal-vacuum testing. In this case, only the bulk solar wind collectors are deployed; arrays for collecting low-speed, high-speed, and coronal mass ejection (CME) solar wind are stowed below the bulk solar wind collection array. Symbols: (a) stack of deployable collector arrays – only the top of the stack, a bulk solar wind collector, is visible; (b) concentrator target holder – actual target is located just below the sunshade; (c) canister-cover collector array; (d) metallic glass collector covering the top of the array-deployment mechanism; (e) gold foil mounted on a side collector; (f) polished aluminum side collector; (g) molybdenum-coated foils in the lid of the sample-return capsule.

to the solar wind) independently when specific types ('regimes') of solar wind are present, thereby providing independent samples of solar wind from different regimes (Burnett *et al.*, 2003). Between exposures, the deployable collector arrays will be shaded by the uppermost, bulk-solar-wind collection array. Note that the deployable arrays can be seen in Figure 2 (a side view of the stack of collector arrays) and Figure 3 (the deployed arrays on the spacecraft) but not in Figure 1 (a top view with the deployable-arrays stowed).

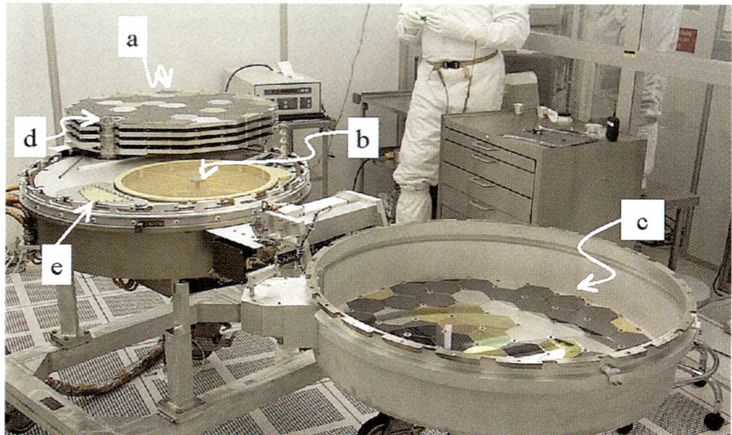

Figure 2. Genesis sample-return capsule, opened during testing after final assembly in the Class 10 clean room in the Johnson Space Center Curatorial Facility. The stack of collector arrays can be seen in side-view. Collector-array materials are installed separately as individual hexagonal (or half-hexagon) wafers. Quality-assurance personnel shown for scale. Symbols as in Figure 1.

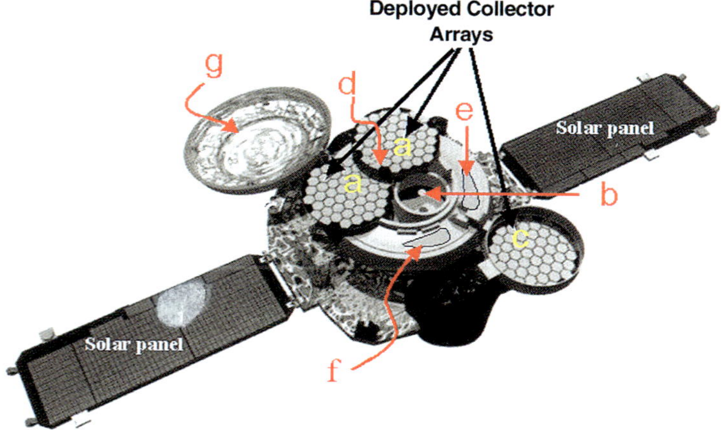

Figure 3. Cartoon of the Genesis spacecraft in collection configuration. Symbols as in Figure 1.

The fifteen materials selected for solar wind collection and their positions are listed in Table I, and an overview for each material is given in Section 3. A discussion of how they were selected is given below.

2. The Material Selection Process

The process of choosing the materials was complex, as the criteria used in the selection were influenced by a variety of factors, some of which are not necessarily intuitive. For example, different techniques for manufacturing otherwise equivalent, ultra-pure materials incorporate different trace impurities. So, manufacturing technique entered into consideration (cf., float-zone silicon vs. Czochralski-pulled silicon vs. vapor-deposited silicon; Wiens and Burnett, 1999). Similarly, solar-thermal properties such as emissivity and solar absorbance were important because the amount of diffusive-loss during collection will depend upon the collector's ambient temperature (i.e., solar-heating). Bulk purity, surface and interface cleanliness, diffusion coefficients (retentiveness) and analyzability had to be weighted with respect to specific elements that the material was a candidate to collect.

Fortunately, data for many of the selection parameters are in the semiconductor industry literature and could be used as a baseline. However, the final selection was based on our own testing whenever time allowed. In other cases (e.g., diffusion coefficients for specific elements and/or specialty collector materials) there was no choice but to determine parameters through *in house* experiments by members of the Genesis science team. Prior to assembly, the science team focused on those materials that carried out the highest-priority measurement objectives required by the Genesis Science Requirements (Burnett *et al.*, 2003). Now, after launch, the science team is still working, determining the usefulness of specific collectors for elements in addition to the highest-priority goals. Before the actual sample return, these details will be published on a Genesis website as a report available to the planetary materials community-at-large. The website is envisioned as a document that can be updated as new information is obtained, and as a resource for future researchers. This concept is already being tested by the Genesis plasma data website (genesis.lanl.gov), which currently has a link to the older Genesis material's selection and feasibility information, www.gps.caltech.edu/genesis, (Wiens and Burnett, 1999) and will eventually have a link to the most current information on the Genesis collector materials (both collector arrays and concentrator targets). Prior to the completion of links to updated materials, it is suggested that researchers requiring specific details on Genesis materials contact the second author (D. S. Burnett).

The Genesis materials-selection criteria are discussed below, along with examples of how they were applied.

2.1. Ultra-pure Materials as Collectors

Purity was the prime science requirement for the Genesis collectors because, in an ultra-pure material, an analytical measurement should give a large signal-to-noise ratio (necessary for high analytical precision) for even small additions of specific solar-wind components.

For elements and isotopes, Genesis Mission purity requirement was determined by error analysis. The requirement was that the background-impurity concentration for a given element be 10% or less than the expected contribution from the solar wind during $\sim$ 2 years of exposure (Burnett *et al.*, 2003, Table II). Where practicable, our goal was to have impurity levels at 1% or less of the solar wind. Therefore,

TABLE I

Suite of materials chosen to be solar-wind collectors, their position in the Genesis sample-return canister, and their intended use. Designated collector materials may eventually be useful for other elements. Moreover, all relevant portions of the sample return capsule will be archived for future solar-wind, micrometeorite, and other studies, as per the Apollo tradition. Position (symbol): collector arrays (CA); concentrator target (CT); array-deployment cover (ADC); side-collector (SC); sample-return capsule lid (SRCL). A summary of the suite of materials in each array (in percent) is given in Figure 5.

Collector Material	Wafer/collector type	Position	Target elements/Intended use
^{13}C Diamond*	Thick film: isotopically enriched diamond grown on commercial product	CT	O, N, F.
Silicon carbide	Undoped ('intrinsic') commercial product	CT	O, N, Li, Be, B, F
Diamond-like carbon[a]	Thick film: Diamond-like carbon on silicon*	CA, CT	N, noble gas isotopes
Aluminum	Laminate: aluminum on sapphire**	CA	Noble gases
Silicon	Czochralski-grown silicon	CA	Same as FZ, except for C and O. CZ is used because it can be obtained with very clean surfaces.
	Float-zone silicon	CA	FZ silicon is exceptionally pure. As far as we know, all elements except those which diffuse rapidly (Fe, alkalis) could be analyzed.
	Silicon on sapphire	CA	C; epitaxial-silicon layer potentially simplifies extraction.
Germanium		CA	No detectable impurities. Ge complements Si for SIMS analysis because of greatly reduced molecular interferences.
Gold**	Laminate: Gold on sapphire**	CA	N, Fe, alkalis.
Sapphire		CA	We are unable to detect any impurities. Potentially useful for alkalis
Carbon+cobalt+gold on sapphire[b]	Laminate separating ions of differing energies	CA	Layered film; SEP-particles
Bulk metallic glass[a]		ADC	Noble gases; SEP-particles
Aluminum alloy	polished Al6061	SC	Bulk solar wind
Gold	Gold foil	SC	N, bulk solar wind
Molybdenum[b]	Laminate: Molybdenum on Platinum**	SRCL	Radionuclides (e.g., ^{10}Be, ^{14}C)

[a]Indicates a specialty (bulk) material.

[b]Indicates a specialty thin film(s) on a commercial product; unmarked indicates materials commercially available, although the shape was a special order

TABLE II

Ambient collector temperatures for representative materials under solar-exposure conditions. These temperatures were measured during the thermal-vacuum test on the engineering-model and depend on the optical properties of the material. Since the concentrator-target materials face away from the sun, their ambient temperature is determined by the optical properties of the fixture.

Concentrator target materials		Temperature (°C)
	Silicon carbide	140
	^{13}C diamond	140
	Diamond-like carbon on silicon	140
Collector array materials		
	Float-zone silicon	141
	Czochralski-grown silicon	156
	Sapphire	56
	Aluminum on sapphire	130
	Gold on sapphire	140
	Germanium	162
Sample-return capsule Lid foil	Molybdenum on platinum	≤ 250

for each solar-wind element, determining whether or not the potential collector material was pure enough required: (a) analyzing that element in the bulk material, and then (b) comparing the measured value with a solar-wind concentration calculated from the estimated fluence implanted by the solar wind during the collector's $\sim$ 2 year exposure[1]. To convert fluence to concentration, it was assumed that the solar-wind fluence was distributed over the outer 1000 Å of the collector.

Potential collector materials were sent to academic and commercial laboratories around the country for trace-element analysis. Often, measured concentrations were at or below the detection limit of the technique. However, in many cases these detection limits implied upper limits to element concentrations that were lower than the 10% solar wind science requirement. Therefore, these materials were still of acceptable purity. Future measurements with improved instrumentation will yield background values below our current baseline.

A summary of the laboratory results for elemental purity: at launch, we knew that (1) we had materials pure enough to collect and measure 44 elements; (2) our analytical sensitivity was insufficient to test purity for the remaining 37 elements; and, (3) there is no element for which we know that there is no collector material

[1] Anticipated solar-wind fluences (atoms cm^{-2}) were based on the solar abundance estimates of Anders and Grevesse (1989), and an average solar-wind proton flux of $\sim 3 \times 10^{+8}$ cm^{-2} s^{-1} (Burnett *et al.*, 2002).

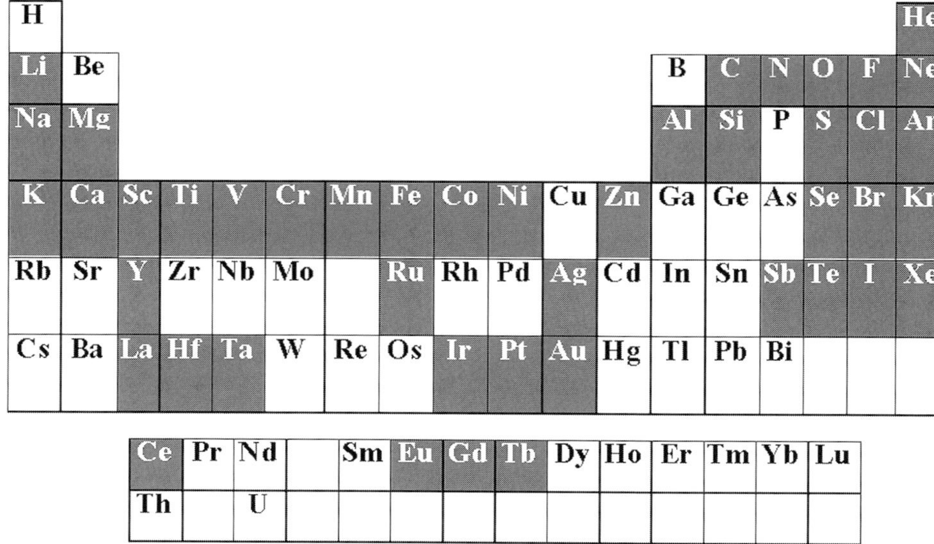

Figure 4. Periodic table of the elements giving an integrated purity overview. We have not yet found an element that is not potentially analyzable in at least one of the materials. White lettering on gray background: elements known to have collector materials of bulk purity sufficient for elemental analysis at the time of the Genesis launch. Black lettering on white background: it is not yet known what collector material(s), if any, will be used for that element.

of adequate purity. The distribution of elements with undocumented purity with respect to their location on the periodic table is given in Figure 4.

2.2. Analyzability of Collectors

Although purity was a prime requirement for materials selection, there is an obvious risk in using even an infinitely pure material for which there is no potential scheme for analysis. Therefore, if a given technique for analyzing a given element or isotope was known to have an especially high analytical precision and sensitivity, then materials easily analyzed by that technique were given preference. Analyzability was not a firm pass/fail criteria, however, since the Genesis mission mandates the future development of high-precision analytical instrumentation.

In the selection of materials for isotopic analyses, analyzability played a major role, especially for oxygen. The literature suggested that float-zone silicon was sufficiently pure that oxygen isotopes could be measured after a simple, two-year exposure to the solar wind. However, high-precision analysis would require a large collection area; moreover, silicon forms a native oxide layer that would interfere with analyses unless removed or otherwise avoided. To avoid these practical problems, the 'concentrator' was developed. This instrument is an electrostatic mirror with materials collecting solar wind for isotopic analysis placed at the focal point. It concentrates elements in the 4 to 28 amu range, increasing the 2-year fluence of these isotopes by an average factor of $\sim 20\times$ (Nordholt *et al.*, 2003; Wiens *et al.*, 2003).

Even with the design and development of the solar-wind concentrator, there was still some concern about quantitatively analyzing the relatively scarce ^{17}O isotope. The science team decided to make at least one concentrator-target either diamond or diamond-like carbon so that CO mass spectrometry could be used (Butterworth *et al.*, 2000). A problem was that, normally, CO mass spectrometry for the 3-isotopes of oxygen (^{16}O, ^{17}O, ^{18}O) uses m/z 28:29:30. But, diamond and diamond-like carbon almost invariably contain at least some terrestrial nitrogen impurities. Therefore, to calculate for ^{17}O we have to correct for background at m/z 28 and 29 ($^{14}N^{14}N$, $^{12}C^{16}O$ and $^{15}N^{14}N$, $^{13}C^{16}O$, plus some cracking from organics). By using ^{13}C in place of normal ^{12}C in the target material, we shift the 3-isotope spectrum up 1amu to m/z 29:30:31. Accordingly, there are two main advantages: the shift separates measured solar wind atoms from terrestrial nitrogen contamination, and strips away almost all isobaric interference (e.g., $^{13}C^{16}O$ interference with $^{12}C^{17}O$) from the $^{17}O/^{18}O$ measurement. These improvements lead to at least a 5-fold improvement in signal-to-background, extremely important given the sensitivity requirements for Genesis.

A more mundane way of making solar-wind elements more analyzable in the collectors array was to decrease the volume of collector material used to catch the 2-year solar-wind fluence. That is, consider an ideal collector that consists only of a clean, high purity, extremely thin membrane: then, the number of implanted solar-wind ions may be large relative to contaminants in the collector simply because there is so little mass in the collector itself. This ideal scenario is equivalent to a layered collector in which a thin layer of collector material is deposited on a clean, inert substrate that provides structural support only. On Genesis, this concept is represented by using high-purity, thin-films deposited on sapphire. Thin metallic films on sapphire can be analyzed separately from the sapphire itself through either laser ablation or differential dissolution.

2.3. The Role of Surface and Interface Cleanliness

Because the solar wind will be implanted near the surface ($\leq\sim$ 1000 Å, depending on the material), being able to differentiate between surface contamination and the implanted solar wind is a significant science issue. The limits for terrestrially-derived surface contaminants are stipulated in the Genesis Mission science requirements. Specifically, at the time of analysis, the surface concentrations of C, N, and O must be $< 10^{15}$ atoms cm^{-2}; for the other elements, the terrestrial contamination must be lower than the estimated solar-wind fluence of the element of interest. These limits were based on the experience of the science team that, in practice, surface cleaning could decontaminate surfaces by no more than a factor of 10–100. Thus, if the contaminant atom/cm^2 were both equal to the solar-wind fluence and could be cleaned by at least a factor of 10, then the minimum purity requirements could be met at the time of analysis. Additional signal-to-background enhancement could then be achieved if analytical techniques with good depth resolution are employed.

For solar wind measurements, analytical techniques are required that are depth sensitive. That is, the techniques need to independently analyze the implanted zone, distinguishing it from both the bulk of the collector material (deeper than the implantation zone) and from any residual surface contamination layer. For example, surface-sensitive techniques such as Secondary Ion Mass Spectroscopy (SIMS) will be able to distinguish between the implanted solar wind and surface contamination. The requirements are still necessary, however, because during SIMS analysis with a high impact-energy primary ion beam (e.g., 20 keV Cs), some fraction of the surface-contaminant atoms ($\sim 1\%$) can be driven to depths where they would interfere with the analysis of implanted solar-wind ($\leq\sim$ 1000 Å) by atomic knock-on effects. Therefore, if there were no requirements on the surface contamination, a component of terrestrial surface contamination at or above the solar-wind fluence could be the limiting factor on the analytical precision during SIMS analysis of a collector. Defining surface-cleanliness standards for Genesis, it was conservatively assumed that stray elemental counts from the surface are $\sim 1-10\%$ of the contamination.

In order to meet the stringent cleanliness requirements, we concluded that it was not possible to defer cleaning until after recovery. Rather, it was necessary to install clean materials in the canister and to keep them clean throughout the duration of the mission. Consequently, all collectors were installed under the cleanest possible conditions (cf., Figure 2) and then enclosed inside the Genesis canister while still in the Genesis Class 10 curatorial cleanrooms at JSC in order to keep them as clean as possible. The exceptions are the foils that line the lid of the sample-return capsule: these were installed in the high bay at Lockheed-Martin Astronautics. These 'lid-foil' collectors (Figures 1 and 3) are designated for collecting radioactive nuclei, and were a special case because the potential for external contamination by radioactive nuclei was considered negligible.

The required levels of surface cleanliness can sometimes be obtained by cleaning procedures and maintained through handling techniques (e.g., Stansbery *et al.*, 2001). Semiconductor industry standards are very high and many materials, e.g. silicon for the collector-arrays, were used as received. For other materials, optical inspection, optical microscopy and/or surface-analysis techniques (e.g., X-ray Photo-electron Spectrometry, XPS) were performed prior to installation in the payload canister. For the concentrator target material, semiconductor cleaning techniques were used to remove particles and/or surface films. Similarly, the specialty laminated materials gold on sapphire and aluminum on sapphire were fabricated using a hot-stage under vacuum, so that volatile adsorbed species could be desorbed before coating. In many cases, witness coupons (e.g., fragments of silicon, gold-on-sapphire, or other material of known surface cleanliness) followed collector materials though the fabrication and installation process, so that surface contamination could be evaluated after the fact by analysis of the coupon.

There is still the possibility of removing some contaminants at the time of analysis, if necessary. For example, adsorbed volatiles might be removed through heating prior to analysis. Moreover, both our experience and that of colleagues in the semiconductor industry suggest that there are cleaning procedures developed for semiconductors that will not remove a significant fraction of the implanted solar wind in our collectors (i.e., that will remove less than the outer $\sim$ 50 Å of the collector), such as low-energy ion milling and low-density plasma etching.

2.4. Testing of Physical Properties

Because Genesis uses passive collectors, the only physical/structural criterion was that the materials used for collection would withstand the launch, flight and re-entry environments. To qualify for flight, individual materials had to undergo a series of environmental tests. Random samples of candidate hexagonal wafers for the collector arrays underwent 3-axis vibration in groups of three wafers; concentrator-target candidates also underwent a 3-axis vibration test in flight-like fixtures. Laminated wafers intended for collector arrays underwent thermal cycling (six-cycles at 20 °C to −170 °C). In addition, gold on sapphire (AuOS) and aluminum on sapphire (AlOS) went through tape-testing – pealing strips of tape from the surface of the laminate – to check the adhesion of the specialty films as well as additional months of thermal-fatigue testing (daily thermal cycling between room temperature and 0 °C). Inside an engineering-model of the canister, all materials underwent both another 3-axis vibration test simulating launch, and a subsequent thermal-vacuum test to simulate the environment expected in space. Following assembly, the Genesis spacecraft itself underwent vibration and thermal-vacuum testing.

After each of these tests, the wafers were inspected for signs of catastrophic failure such as cracking and chipping. In terms of influencing the selection, the impact was generally to modify the specifications for the collectors, rather than the choice of material itself. For example, the tape testing of AuOS and AlOS indicated that sapphire wafers needed to be cleaned in an O_2 plasma after etching in order to obtain the maximum adherence between the film and the substrate. Similarly, the thickness of all of the collectors in the canister-cover collector array was changed from 500 μ to 700 μ, because 500 μ silicon collectors had cracked during 3-wafer vibration tests, and because cracks or chips were observed in other 500 μ collectors, including silicon and silicon-on sapphire, after the vibration test of the engineering-model canister.

2.5. Solar-Thermal Properties of Collector Materials

The ability of a material to retain solar wind is clearly an important factor in the selection. Each collector material will retain some solar-wind components more efficiently than others because of differences in chemical and structural properties.

The optical properties of materials determine their ambient temperature during exposure to the sun; therefore, in general, the optical properties of individual collector materials had to be determined to calculate the ambient collection temperature and its influence on the diffusion rates of solar-wind components. Because the concentrator target materials are inverted (facing the focal point of the electrostatic mirror) their exposure temperature is fixed by the optical properties of their holder.

Genesis engineers modeled the temperatures to be expected during solar-wind collection using emissivity and solar absorbance values measured at the Jet Propulsion Laboratory/California Institute of Technology (JPL). Emissivity measurements were performed using a Gier Dunkle Infrared Reflectometer (Model DB100). Solar absorbance measurements were performed using a Gier Dunkle Reflectometer (Model MS251). In addition, during thermal-vacuum testing, the engineering-model canister was wired with thermocouples in order to measure collector temperatures directly. Table II gives representative collector material temperatures measured inside of the canister during the solar-thermal vacuum test of the engineering model. We believe that these measured temperatures will closely approximate the ambient temperatures that individual collectors will experience during the mission.

We note that in most cases the ambient temperatures of the collectors are relatively low and will not dramatically enhance diffusive loss of implanted solar-wind components. There are also advantages to having the collectors warm, but not hot. For example, contaminants that outgas from the warm spacecraft will condense at cold points rather than sticking to the warm collectors (Wiens and Burnett, 1999). In metallic collector materials, heating may mitigate radiation damage caused by the solar wind through annealing and the stabilization of defects.

2.6. Elemental diffusion of solar-wind in collector materials

The potential for the steady-state diffusive-loss of solar-wind elements from the material during collection was a major criterion in selecting collector materials. Estimating this loss required knowledge of two components: the ambient temperature that the collector would reach during exposure to the sun, and the diffusion coefficient for that element in the collector at that ambient temperature. When both were known, diffusion rates for solar-wind elements in candidate collector materials could be estimated.

For materials selection and testing, we allowed for ambient temperatures up to 200 °C, which was a conservatively high estimate from what was measured during testing (cf., Table II, and Section 2.5). In some cases, diffusion coefficients could be gleaned from the literature; however, if no appropriate literature value was available, diffusion rates were bounded *in house* using actual diffusion experiments performed on implanted candidate materials (e.g., Wiens and Burnett, 1999; Jurewicz *et al.*, 2000).

The variety of materials on the collector arrays is a good example of how retention factored into the selection of collector materials. The semiconductor literature indicated that alkali elements (at least Li, Na, and K) and Fe will diffuse over large distances in silicon, and similar results were determined *in house* for Ne in silicon (Wiens and Burnett, 1999). Accordingly, other collector materials (e.g., sapphire, gold on sapphire) were added to cover these elements.

In the cases of the specialty materials used on Genesis, clearly no literature values were available. Accordingly, diffusion rates were bounded using actual diffusion experiments performed on implanted samples of these candidate materials (e.g., Meshik *et al.*, 2000).

Researchers requiring diffusion information can currently find tabulated results from the original Genesis feasibility study and proposal in Wiens and Burnett (1999) electronically, through a link on the Genesis plasma data website (genesis.lanl.gov). Details of the retentivity of solar-wind components in collector materials are still being compiled. Data will eventually be available in final reports; meanwhile pressing questions on information not currently on the web can be discussed with the second author (D. S. Burnett).

3. Flight Solar-wind Collection Materials

The Genesis spacecraft is flying fifteen individual materials designated for solar-wind collection. Again, these collector materials are distributed among (1) the target of the concentrator (Nordholt *et al.*, 2003; Wiens *et al.*, 2003), (2) the collector arrays and (3) the supplemental collectors, e.g., two side collectors, the cap on the array-deployment mechanism, and the sample-return capsule lid foil (Figures 1–3; cf., Table I). A discussion of the individual materials organized by position on the spacecraft is given below.

3.1. Concentrator Target Materials

At the focal point of the solar-wind concentrator is a target that is approximately 6 cm in diameter, giving a collection area of $\sim$ 25 cm^2 of material (Nordholt *et al.*, 2003; Wiens *et al.*, 2003). The target holder is divided into four quadrants each containing a material tailored to collect light elements whose isotopes are of fundamental scientific interest. Of the four quadrants: one is an amorphous diamond-like carbon, one is ^{13}C-diamond, and two are silicon carbide. Because the concentrator separates positive ions from electrons, these collector materials have an additional selection criteria in that they must have a surface resistivity of less than $\sim 10^{11}$ ohm-cm. The amorphous carbon and the silicon carbide had sufficient conductivity; but, the diamond had to be coated to meet this criterion. At JPL, the flight quadrants were cleaned and an XPS analysis of each quadrant was performed just prior to the final particulate-removal process to insure surface cleanliness. Installation was performed at Los Alamos National Laboratory (LANL).

Detailed chemical properties used in the materials selection will be given in the Genesis concentrator-target materials final report (D. S. Burnett, in preparation).

TABLE III

A comparison of the purity and surface-contamination concentrator target materials with data for the Genesis science requirements for the top two Genesis mission priorities: measurement of oxygen and nitrogen isotopes.

Material	Purity (% expected solar wind)		Surface contamination (atoms/cm^2)	
	O	N	O	N
Silicon carbide	0.1	0.5	8e14	<5e14
^{13}C diamond	<1	8	3e15	<5e14
Sandia diamond-like carbon	2	1	3e15	<5e14
Science requirement[a]	10	10	1e15	1e15

[a]The requirement for surface contamination is defined at the time of analysis and, therefore, allows for post-flight cleaning if necessary.

Meanwhile, a summary of how well these materials meet the purity and surface contamination for the top two Genesis Mission science priorities (measurement of O and N isotopes) is given in Table III, and a general overview of the materials themselves is given below.

3.1.1. *Diamond-Like Carbon on Silicon*

This specialty material is targeting nitrogen and its isotopes, because N_2 can be readily extracted for gas-source mass spectrometry using stepwise oxidation of the carbon. Although potentially useful for oxygen, the background levels are significantly higher than for either the ^{13}C diamond or the silicon carbide targets.

The diamond-like carbon is a thick (1.5 μ–3 μ), semi-ordered, film on a silicon substrate that was made for Genesis by Sandia National Laboratories (Friedmann *et al.*, 1994; Alam *et al.*, 2001). Using pulsed laser deposition, carbon was sputtered from a target onto a silicon substrate under vacuum. To prevent delamination within the finished, thick film, *in situ* annealing steps were performed after each (thin) deposition to relieve that layer's internal stresses. Because of equilibration with the chamber atmosphere during these annealing steps, there is an internal, layered distribution of terrestrial contaminants (O, F, lesser amounts of N) every 1000–1500 Å throughout the standard material. For Genesis, in order to avoid internal contamination in the zone destined to collect solar wind, the second to last annealing step was skipped; accordingly, the upper 2000–3000 Å of the quadrant is not layered with these terrestrial contaminants, but the surface contamination is relatively high (Table III).

This material was also chosen for the collector arrays and will be deployed as full hexagons (3 each on the bulk solar-wind collector arrays, 4 on each of the three other arrays). However, the special annealing cycle described above for the concentrator was not used. In addition, a 50 Å, ^{13}C version of this material was used as the conductive coating on the ^{13}C diamond concentrator target.

3.1.2. ^{13}C *Diamond*

Thick, freestanding crystalline-diamond films are produced commercially by several vendors, many of whom use proprietary processes. Examples of non-proprietary processes are given by US Patents #5,273,731 and #5,110,579 assigned to General Electric Corp in 1993 and 1989, respectively (Anthony and Fleischer, 1993, Anthony and Fleischer, 1992). Genesis required a specialty diamond product in which the collection surface was a diamond composed of the isotope ^{13}C, so that mass spectrometric analysis of oxygen isotopes through conversion of oxygen-implanted diamond to CO would have minimal interferences for the $C^{17}O$ species even in the presence of minor amounts of terrestrial nitrogen impurities (e.g., $^{13}C^{17}O$ is not affected by interference with $^{13}C^{16}O$ or $^{14}N^{15}N$ as $^{12}C^{17}O$ would be, and $^{12}C^{18}O$ interference will be negligible with a ^{13}C substrate; cf., Section 2.2, Analyzability of collectors). Fluorine and nitrogen should also be analyzable, so the diamond can be used for other than oxygen if enough material is available.

Raytheon was the vendor chosen to make this material both because of the purity of their standard diamond films and their willingness to develop process techniques for manufacturing the specialty material. Because of the high cost of isotopically-enriched methane, only the top 50 μ or so is the specialty diamond; the rest of the 570 μ thick wafer is the standard commercial material. Wafers were laser cut to shape, mechanically ground, and polished. Between the final cleaning at JPL and installation at LANL, the quadrant was sent to Sandia National Laboratory, where a 50 Å ^{13}C diamond-like carbon layer (Alam *et al.*, 2001) was deposited on the surface, in order meet the conductivity requirement. The diamond-like carbon film may need to be removed before laboratory analysis because it is higher in terrestrial atmospheric components than the underlying diamond.

3.1.3. *Silicon Carbide*

This commercial material (purchased from CREE, Inc.) comprises 2 quadrants of the 4 concentrator targets, $\sim 12\ cm^2$ of collection material. The collection surface is an ultra-high-purity SiC layer (~ 2 microns thick) grown epitaxially on a nitrogen-doped, polished silicon-carbide wafer. ('Epitaxial' indicates that the coating shares a crystal-lattice plane with the substrate, and is single-crystal). CREE, Inc.'s fabrication technique is proprietary; however, their website, www.cree.com, currently lists a number of references for the epitaxial growth of silicon-carbide films, many of which are variations of either molecular-beam epitaxy or chemical vapor deposition (e.g., Fissel *et al.*, 1995; Powell *et al.*, 1991). After the epitaxy, CREE then laser-cut the resultant wafer to shape. The SiC is sufficiently conductive that no coating was necessary. To meet the structural criteria for flight, a second vendor (RhioTech) mechanically ground and shaped the edges. It is probable that oxygen, nitrogen, fluorine, lithium, beryllium, and boron will be analyzable in this material.

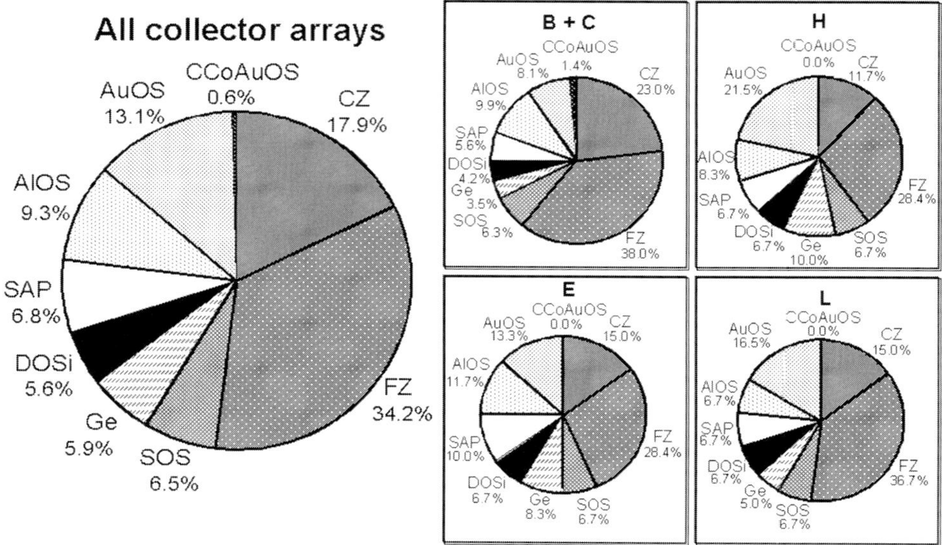

Figure 5. Distribution of collector materials located in the Genesis collector arrays, in percent. Notice that each array contains between ~50% and 60% silicon. Materials designations: (CZ) Czochralski-grown silicon; (FZ) Float-zone silicon; (SOS) silicon on sapphire; (Ge) germanium; (DOSi) diamond-like carbon on silicon; (SAP) sapphire; (AlOS) aluminum on sapphire; (AuOS) gold on sapphire; (CCoAuOS) carbon on cobalt on gold on sapphire. Collector array designations (Bold Capital letters): (B+C) the two collector arrays that capture the bulk solar wind, where B is the top of the stack of collector arrays and C is the inside of the canister cover; (E) collector array capturing the solar wind from coronal mass ejections; (L) low-speed solar wind collector array; (H) high-speed solar wind collector array.

3.2. Collector-Array Materials

The bulk of the material for collecting solar wind is found in the Genesis collector arrays. Each array holds 54 (55 in the canister-cover array) four-inch diameter hexagonal collectors and 6 collectors shaped as half-hexagons. The distribution of the materials within each array is given in Figure 5. Detailed chemical properties used in their selection will eventually be given on a Genesis website as a Collector-Array Materials Final Report (D. S. Burnett, in preparation). Meanwhile, some older information is available in Wiens and Burnett (1999). In all cases, flight-like materials have been archived at JSC as control samples. For tracing impurities in collectors cut from silicon boules (large, crystollographically-oriented, synthetic single crystals, such as made by Czochralski- and Float-zone growth processes), the terminations of the boules were cut and archived, and the approximate location along the length of the crystal from which the individual collector was cut was recorded. An overview of each collector material is given below.

3.2.1. *Silicon*

Because we have certification of high-purity and a plethora of information from the semiconductor literature on other selection criteria, commercial silicon constitutes more than half of the hexagons in the arrays. There are three individual types of silicon made using three significantly different fabrication techniques: Czochralski-grown silicon, float-zone silicon, and silicon on sapphire (O'Mara *et al.*, 1990, for details; overview below). The three manufacturing techniques result in different trace impurities. The Czochralski-grown wafers and the float-zone wafers are solid silicon; silicon-on-sapphire wafers consist of ~ 3000 Å of silicon grown epitaxially on a hexagonal sapphire substrate.

3.2.1.1. *Czochralski-grown silicon.* Czochralski growth is where a small seed crystal is slowly pulled from a melt to produce a large, single-crystal boule. For silicon made by this process, there is some potential for the melt to leach contaminants from the crucible holding the silicon melt, and then incorporate them into the boule. However, manufacturing procedures (as well as the large volumes manufactured) insure fabrication of high purity silicon.

For Genesis, the silicon boules to be cut into collectors were a commercial semi-conductor product purchased from MEMC Electronics (MEMC). All of the Czochralski wafers were cut from two single-crystal boules grown for Genesis, and were tracked with respect to both the crystal, as well as the section of the crystal, from which the collector was cut. Each small group of wafers was given a serial number. As quality control, cores were bored from slabs cut from the terminations of the crystals and were sent out for analysis to determine minimum bulk purity.

These round, commercial wafers were then cut into hexagons using laser cutting (Directed Light), mechanically edge-ground to remove flaws and increase their physical strength (RioTech), and then returned to MEMC for cleaning. In order to be flight-quality, these collectors had to have stringent cleanliness and quality control; e.g., 5 or fewer particles $< 0.5\ \mu$, and no scratches, stains or solution-spots on the collection surface.

3.2.1.2. *Float-zone grown silicon.* Float-zone growth is where a high-purity but multi-granular rod is melted and recrystallizes to a single crystal when passed through an RF-heating coil. The melt never touches a crucible; moreover, as the rod melts *in situ*, the impurities in the already pure rod are zone-refined (tend to stay in the melt, not the crystal) and are effectively eliminated from the newly-grown bulk crystal. Accordingly, float-zone grown materials have a reputation for ultra-high purity.

Single-crystal semiconductor-grade silicon made by this method is a standard commercial product. The single-metal crystals are grown under an argon atmosphere to prevent oxidation. This method of fabrication produces material of higher purity than the Czochralski-pulling technique, including a lower C, N, and O content.

In the semiconductor industry impurities imply conduction via defects; therefore, the resistivity of a wafer is used as a general indicator of overall purity. Float-zone silicon wafers grown from normal stock typically have impurity levels of ~ 2000 ohm cm; those produced from ultra-high purity poly-silicon average ~ 18 000 ohm cm. We have obtained material with over 50 000 ohm cm.

In a procedure similar to that used for the Czochralski-grown collectors, round commercial wafers were purchased from the vendor (Unisil), laser-cut into hexagons (Directed Light), edge-ground to meet flight specifications (RhioTech), and then cleaned to flight specifications (MEMC).

3.2.1.3. *Silicon on sapphire.* This laminate was produced by Kyocera Industrial Ceramics Corporation (Kyocera), and is nonstandard only in it's hexagonal geometry. A 3000 Å silicon layer is epitaxially grown on a pre-formed, polished, high-quality single-crystal sapphire substrate using a proprietary process. This finished collector was then shipped directly to JSC for a fit-check and installation into the collector arrays, so there was no intermediate handling. The material is clean under XPS analysis, but no other data exist at present on surface contamination levels. Unfortunately, since procurement of collectors for the Genesis mission, Kyocera has discontinued its silicon-on-sapphire product line. Flight-like wafers are, however, archived at JSC for future analysis.

In this case, the thin silicon layer is the solar-wind collector and the sapphire is merely a support. Sapphire is desirable as a substrate because it is transparent to many laser wavelengths and is relatively inert chemically. Therefore, this material was originally included to aid analytical techniques using laser-ablation and/or differential dissolution. However, because the silicon is epitaxial and shares a crystallographic orientation with the substrate, the silicon is also single crystal. Because of this crystallographic coherence between the silicon and the sapphire, it proved to be tougher than either silicon or sapphire alone for engineering purposes. As a solar-wind collector, recent work is proving that the silicon layer has adequate purity for a number of elements (Humayan *et al.*, 2001).

3.2.2. *Sapphire*

Throughout the commercial sector, high-purity, single-crystal Al_2O_3 is known as 'sapphire'; this is in contrast to the jargon of the geological community, who would refer to this material mineralogically as high-purity corundum. The sapphire used by Genesis is a commercial product purchased from Kyocera. For electronics, sapphire is usually used as an insulating substrate; for Genesis, as a collector it is used to collect and retain alkalis in the solar wind, as diffusion rates for alkalis in silicon are known to be relatively high (Treiman, 1993).

The sapphire is single-crystal grown by the edge-defined film-fed growth method. A general description of this technique can be found on the Kyocera website and literature (cf., www.kyocera.com/KICC/industrial/products/crystal.htm); a more technical overview can be found in Scott *et al.* (1995). This crystal-growth technique allows Kyocera to make wafers in unusual geometries without the need for excessive cutting and grinding. Potential manufacturing defects include bubbles or other inhomogenieties near the edge of the wafer; however, inspection did not reveal any defects in the Genesis collectors.

Preliminary SIMS analyses by science team members showed that the sapphire easily met bulk purity criteria for Na, K and Si. The fact that the sapphire is insulating required that it be coated before both SIMS analysis and successful ion-implantation performed for laboratory experiments. However, charging is not an issue for the collector array materials because the solar wind is a plasma, and the amount of surface charging during collection has been calculated to be on the order of 10 V.

3.2.3. *Germanium*

These collectors of semiconductor-grade material were brokered through International Wafer Service. Boules of single-crystal germanium were cut, shaped and polished by their subcontractors. Although they are physically more fragile than silicon, no breakage of collectors was observed during the spacecraft's vibration test.

Germanium was chosen for the arrays primarily because (1) the high mass of germanium eliminates low-mass molecular interferences in trace-element analysis techniques that are based upon mass-spectroscopy, and (2) semi-conductor germanium is reputed to be of extremely high bulk purity. However, *in-house* verification of the purity was difficult, and we have little data confirming purity for this material. Accordingly, only $\sim 6\%$ of the wafers in the collector arrays are germanium.

3.2.4. *Specialty Laminate Materials*

Approximately a quarter of the collectors in the arrays are specialty laminate materials prepared either at Sandia National Laboratory (diamond-like carbon on silicon) or in conjunction with JPL (gold on sapphire; aluminum on sapphire; carbon on cobalt on gold on sapphire). In all cases, the silicon or sapphire substrate is simply there to support the films so that a thick, strong hexagonal wafer can be bolted into the collector arrays; only the film actually collects the solar wind.

Because these materials were made specifically for the Genesis Mission, the Genesis archives contains smaller control samples as well as flight-like wafers. For the diamond-like carbon on silicon, 2 inch diameter control wafers were produced after approximately every third hexagon. For the laminates made under the auspices of JPL, small (~ 2 cm $\times$ 2 cm) witness coupons were included in each of the flight (and flight-like) depositions. These control wafers and witness coupons are archived for future use.

For the aluminum- or gold- on sapphire collectors, there is, at present, effectively no data on surface contamination. However, it was determined by science team members that the interface between the aluminum and sapphire in the (AlOS) could skew a complete noble gas analysis, as will be discussed later.

3.2.4.1. *Diamond-like carbon on silicon.* This material is essentially the same as the concentrator target material discussed previously (Section 3.1.1 and Table III). The only differences are the full-hexagon shape and the fact that all of the annealing steps were performed between layers. Accordingly, there will be some atmospheric contamination in the top 2000 Å.

This material has been added to the collector arrays because it should be retentive and is sufficiently conductive for SIMS analyses even without a coating. The nitrogen contamination from the surface to the outer 1000 Å is about 20% of unconcentrated solar wind. However, if the background is sufficiently uniform, analysis of elemental and isotopic analyses of nitrogen may still be feasible on this material. Development of stepped-combustion extraction techniques would permit noble-gas analyses on this material.

3.2.4.2. *Gold on sapphire.* Gold is a good substrate for SIMS in that molecular ion interferences are negligible. However, gold has the significant disadvantage of backscattering losses during collection, with a resultant isotopic fractionation. The backscatter correction is sufficiently accurate for elemental analysis. But, at present, it is unclear if accuracy of the correction is sufficient for isotopic analysis. For other types of analysis, the gold can be mechanically or chemically removed from the sapphire.

A 2000 Å layer of the highest-purity gold (99.9999%) purchased from ESPI (www.espi-metals.com) was e-beam deposited onto sapphire substrates (Kyocera). Prior to deposition, the substrates were cleaned with solvents, etched in HF, cleaned in an RF-generated O_2-plasma, and placed in the e-beam evaporator within an hour of the O_2-plasma cleaning step. During the pump-down of the e-beam evaporator, the chamber was flushed with argon; then, a heat lamp was turned on to keep the wafers hot overnight. After deposition, the wafers were cooled under vacuum; the chamber was again flushed with argon prior to exposure to air. The chamber used for gold deposition had a better vacuum ($\sim 10^{-8}$ torr) and had been previously used for the evaporation of a more restricted set of elements than that used for aluminum deposition (below).

3.2.4.3. *Aluminum on sapphire.* The intent for making this specialty laminate was to have a thin film of aluminum on a substrate transparent to lasers, to enable laser extraction of noble gases. This material can be used for neon (Meshik *et al.*, 2000) and argon analysis; however, the aluminum-sapphire interface contamination levels appear to be too high for krypton and xenon. We note that silicon on sapphire appears to be an adequate alternate for krypton and xenon, even though the silicon is not sufficiently retentive for neon.

Beyond noble gases, we have no direct measurements of the purity of this material. However, the material was made in the JPL Microdevices Laboratory with bulk purity in mind. The Kyocera substrates were prepared using standard semi-conductor techniques, including cleaning with solvents, etching in HF, final cleaning in an RF-generated O_2-plasma, and then placing the wafers in an e-beam evaporator within an hour of the final cleaning. Then, the vacuum chamber of the e-beam was flushed twice with nitrogen, and the fixture holding the wafers were heated to greater than 220 °C overnight and achieved a base pressure of $\sim\leq 5 \times 10^{-7}$. 'Gettering' (selective removal of reactive gases, in this case oxygen and nitrogen) with chrome and/or titanium further reduced the base pressure ($< 2 \times 10^{-7}$) before the aluminum deposition. 3000 Å of the highest-purity aluminum (99.9999%; Goodfellow) was deposited onto the hot substrates; the e-beam chamber was flushed with nitrogen before removal of the completed collectors in order to mitigate any subsequent oxidation.

3.2.4.4. *Carbon-cobalt-gold on sapphire.* This laminate is part of an attempt to provide suitable material for measuring He and Ne particles with greater than solar-wind energy (SEP). The concept was to have a disposable low-density layer (3500 Å carbon) that would stop the main solar wind, and a thick (1μ), dense layer (gold) to collect the solar energetic particles. The 2500 Å cobalt layer was added in order to mark the interface so that we could insure that all of the carbon had been removed. Surface contamination was considered a secondary issue for this material, because the cobalt layer and/or the carbon layer will be removed before analysis in order to separate the SEP's from normal solar wind. There is only one full-hexagon and two half-hexagons (~ 126 cm^2 of exposed area) on the spacecraft; all are on the B-array (cf., Figure 5), which collects bulk solar wind.

The fabrication process for the gold-on-sapphire portion of the laminate was similar to the fabrication of the gold-on-sapphire collector, except that: (1) the gold used was 99.999% purity, and (2) the gold film was $\sim$ 1 μ thick. The wafers were then transported (in Fluoroware®, under air) from Pasadena, CA to Buellton, CA where cobalt and carbon were sputter-coated under argon by Thin Film Technology, Inc.

3.2.5. *Materials for Other Collectors*

The focus of the Genesis science implementation is on the materials in the concentrator and the collector arrays, because those are the instruments initially designed to attain the highest-priority science objectives. However, there are targets of opportunity for additional collection area (cf., Figures 1 and 2). The top of the array-deployment assembly is exposed to the bulk solar wind whenever the canister is opened. Whenever the collector arrays are deployed, two more areas become exposed to the solar wind. These areas contain a 'canister closeout radiator' that prevents direct sunlight from hitting the bottom of the canister; the top of each radiator contains a 'side collector'. Similarly, the inside of the lid of the sample-return capsule is exposed to the solar wind continuously during collection, so it too is lined with a collector material. The four additional collector materials are: a metallic glass (the array-deployment-assembly cover), gold foil (a side collector riveted to a canister closeout radiator) and a polished aluminum alloy (a side collector comprising the entire top surface of a canister closeout radiator), and molybdenum on platinum (the sample-return canister-lid collector foils).

3.2.5.1. *Bulk metallic glass.* This collector material is an amorphous metal. That is, it is from a new class of multi-component metallic compositions in which the nucleation rate of crystalline metal is greatly suppressed when the liquid is quenched rapidly. These compositions have sluggish crystallization kinetics, enabling them to form 'bulk metallic glasses' (thick pieces of vitreous alloys) at cooling rates comparable to those used for traditional glass formers, such as the sodium-borosilicates that comprise normal window glass. As such, the ability to make vitreous alloys in a variety of compositions has recently triggered a tremendous amount of research into the properties of these unique materials, with the hope of using them in fields ranging from aero-structures to magnetism.

Bulk metallic glasses exhibit properties advantageous for solar-wind collection (Heber, 2002). First, because there is no crystal lattice, fractionation and loss of solar wind due to diffusion along preferred paths on crystal planes is absent. Similarly, there are no grain boundaries to provide high-diffusion paths. Therefore, the glass retains noble gases well, especially He and Ne. Moreover, it etches very uniformly and is, therefore, excellent for analyses that include high-resolution depth profiling (e.g., acid etching; Heber, 2002). This material was included especially to look at the solar energetic particle (SEP) component in the solar wind.

The Genesis Mission wanted a specialty vitreous-alloy composition in order to minimize the potential for contamination in the worst-case scenario in which a micrometeorite impacted the multi-component metallic-glass collector and 'splashed' nearby collectors. The final composition was chosen after researching the retentivity and analyzability of glasses in two composition manifolds: Ti-Zr-Cu-Ni, vitreous alloy 101, and Zr-Nb-Cu-Ni-Al, vitreous alloy 106 (Heber, 2002). The Genesis bulk metallic glass is (in atomic %) is: $Zr_{58.5}Nb_{2.8}Cu_{15.6}Ni_{12.8}Al_{10.3}$.

This new bulk metallic glass was custom-made at the California Institute of Technology. It has been the subject of recent NASA sponsored research via electrostatic levitation processing techniques implemented at the Marshall Space Flight Center (Hays *et al.*, 2001). These measurements, on research grade specimens, have shown that the critical cooling rate required to vitrify the liquid on cooling from the liquid state is 1.75 K s^{-1}. This composition exhibits an excellent thermal stability with respect to crystallization. The difference between the crystallization temperature, T_x, and the glass transition temperature T_g, is $\Delta T = T_x - T_g = 100$ K.

3.2.5.2. *Gold foil.* High-purity gold foil provided by the University of Minnesota was attached to one of the side collectors. The foil is of sufficient purity that nitrogen can be measured in the outer 1000 Å of this material. No other purity or cleanliness data exist at present.

3.2.5.3. *Aluminum metal alloy (polished).* The entire top surface of this radiator is polished to make a collection surface. Thus, the collector is a standard commercial aluminum structural alloy (#6061), is part of a structural unit, and will only be useful if a surface analysis technique is employed. The collection surface was hand-polished by JSC curatorial personnel. Nothing is known at present about the purity or surface cleanliness of this collector; again, it is a structural spacecraft component, not originally intended to be a solar-wind collector. However, it is thought that this 'collector of opportunity' should be adequate for noble gas analysis.

3.2.5.4. *Molybdenum on Platinum.* This specialty suite of solar-wind collectors is part of an experiment to measure radionuclides (e.g., ^{10}Be (Nishiizumi and Caffee, 2001)) produced in the sun. All of the collectors are ($\sim$ 3000 Å) molybdenum on a platinum substrate. Combined, the collectors fill an area of $\sim$ 13 000 cm^2. As with the laminates in the collector arrays, it is actually the molybdenum film that is collecting solar wind, the platinum is merely a chemically-inert structural support.

The collectors were made under the supervision of the science team in the following way. The platinum-foil substrates (purchased from Goodfellow) were first etched by aqua regia to slightly texture the surface in order to modify their optical properties, so that the expected temperature of the foils (when exposed directly to the sun) would be within the spacecraft's engineering tolerance. Then, the foils were coated at a commercial vendor (Thin Film Technology, Inc) during a single deposition by e-beam evaporation of molybdenum. These pieces were hand-cut to shape, sewn onto the multi-layered insulation (MLI blanketing) that was mounted over the parachute deck (inside the lid of the spacecraft's sample-return capsule) at Lockheed Martin Astronautics (LMA). Again, molybdenum was chosen in part for engineering reasons: it's solar-thermal properties indicated it would not overheat the parachute deck. It was also chosen because screening tests showed that it should retain ^{10}Be at the high collection temperatures (over 250 °C) for the two years of collection. Most importantly, solar-flare particles produce negligible amounts of ^{10}Be and ^{14}C in molybdenum, so that the analytical (^{10}Be and ^{14}C) background will not increase with exposure time. Unfortunately, because molybdenum makes a thermodynamically-stable carbide, during the e-beam evaporation it tended to 'getter' carbon from the atmosphere of the large vacuum chamber, thus incorporating it into the molybdenum film. The Genesis science team is currently checking to see if the ^{14}C background is low enough to also measure that isotope.

4. Curation, Allocation and Early Science Return from Genesis Solar-wind Collection Materials

A major goal of Genesis is to provide a reservoir of materials for the scientific community during the 21st Century. Accordingly, the majority of the materials used to collect solar wind on Genesis will be archived at the curatorial facility of the Johnson Space Center. Samples will be allocated to researchers based on recommendations from a Sample Allocation Committee, as per the deep heritage going back to returned Apollo samples. It is planned that the opportunity to analyze these samples will be opened to the international planetary materials community within four months of recovery. No special privileges are reserved for the principal investigator or co-investigators, except for the Early Science Return portion of the mission.

The Early Science Return includes a set of four experiments to be performed by members of the Genesis science team upon retrieval of the sample return capsule: N-isotopes, noble-gas isotopes in bulk solar wind, bulk C-isotopes, and the search for radioactive nuclei. The results of these studies are scheduled for publication in the open literature within 1.5 years of recovery of the Genesis materials. However, these studies will not include any material from the concentrator and will use a small amount of the material from the collector arrays.

Accordingly, all the materials in the concentrator and the vast majority of the wafers from the collector arrays will be available to the planetary materials community at large. Further details are given in the Genesis Mission overview in this issue (Burnett *et al.*, 2003).

Acknowledgements

A portion of this work was carried out through the Jet Propulsion Laboratory, California Institute of Technology, under contract with the National Aeronautics and Space Administration Discovery Program. Special thanks to the commercial vendors who made singular efforts to meet the Genesis requirements, which were both stringent and unusual. Thanks also go to the members of the Genesis Engineering Teams, the JPL MicroDevices Laboratory Central Support Group, the Genesis Project Office, and to Alan Treiman who performed the initial diffusion-stability analysis. Two anonymous reviewers also helped to significantly clarify the content, and contributed their own humor to enhancing the project. Reference herein to any specific commercial product, process, or service by trade name, trademark, manufacturer, or otherwise does not constitute or imply endorsement by the United States Government, or the Jet Propulsion Laboratory, California Institute of Technology.

References

Alam, T. M., Friedmann, T. A., and Jurewicz, A. J. G.: 2001, 'Solid State 13C MAS NMR Investigations of Amorphous Carbon Thin Films: Structural Changes During Annealing', *Thin Films: Preparation, Characterization, and Applications*, Proc., Am. Chem. Soc. Meeting April 1–5, 2001, San Diego, CA.

Anders, E. and Grevesse, N.: 1989, 'Abundances of the Elements: Meteoritic and Solar', *Geochim. Cosmochim. Acta* **53**, 197–214.

Anthony, T. R., and Fleischer, J. F.,: 1992, *Transparent Diamond Films and Method for Making*, General Electric Company, Schenectady NY, assignee. US Pat. #5,110,579.

Anthony, T. R., and Fleischer, J. F.,: 1993, *Substantially Transparent Free Standing Diamond Films*, General Electric Company, Schenectady NY, assignee. US Pat. #5,273,731.

Burnett, D. S., Barraclough, B. L., Bennett, R., Neugebauer, M., Oldham, L. P., Sasaki, C. N., Sevilla, D., Smith, N., Stansbery, E., Sweetnam, D., and Wiens, R. C: 2003, 'The Genesis Discovery Mission: Return of Solar Matter to Earth,' *Space Sci. Rev.*, this volume.

Butterworth, A. L., Franchi, I. A., and Pillinger, C. T.: 2000, 'Solar Wind Sample Return from Genesis: Towards the Extraction and Isotope Ratio Measurement of Nanogram Quantities of Oxygen Implanted into Diamond', LPSC XXXI (cd-rom) 1704.pdf.

Fissel, A., Kaiser, U., Ducke, E., Schroter, B., and Richter, W.: 1995, 'Epitaxial Growth of SiC Thin Films on Si-stabilized SiC (0001) at Low Temperatures by Solid-Source Molecular Beam Epitaxy', *J. Crys. Growth* **154**, 72–80.

Friedmann, T. A., Siegal, M. P., Tallant, D. R., Simpson, R. L., and. Dominguez, R. L.: 1994, in: C. L. Renschler, D. Cox, J. Pouch, and Y. Achiba *Novel Forms of Carbon II*, Materials Research Society, Pittsburg, pp. 501–506.

Hays, C. C., Schroers, J., Johnson, W. L., Rathz, T. J., Hyers, R. W., Rogers, J. R., and Robinson, M. B.; 2001, 'Vitrification and Determination of the Crystallization Time Scales of the Bulk-Metallic-Glass-Forming Liquid $Zr_{58.5}Nb_{2.8}Cu_{15.6}Ni_{12.8}Al_{10.3}$', *Appl. Phys. Letters* **79**, 1605.

Heber, V. S.: 2002, 'Ancient Solar Wind Noble Gases in Lunar and Meteoritic Archives and Tests for Modern Solar Wind Collection with the Genesis Mission', Thesis ETH, Zürich.

Humayun, M., Campbell, A. J., Burnett, D. S., and Jurewicz, A.: 2001:, 'Improving the s/n of HR-ICP/MS: Determining the elemental composition of the solar wind from Nasa's Genesis Spacecraft Mission', Federation of Analytical Chemistry and Spectroscopy Societies. Detroit, MI (October 7–12, 2001)., abst. #**340**.

Jurewicz, A. J. G., Burnett, D. S., Wiens, R. C., and D. Woolum,: 2000, 'Genesis Solar-Wind Sample Return Mission: The Materials', LPSC XXXI (cd-rom) 1783.pdf.

Meshik, A. P., Hohenberg, C. M., Burnett, D. S., Woolum, D. S., and Jurewicz, A. J. G.; 2000, 'Release Profile as an indicator of Solar Wind Neon Loss from Genesis Collectors', *Meteorit. Planetary Sci.* **35**, A109–A109.

Nordholt, J. E., Wiens, R. C., Abeyta, R. A., Baldonado, J. R., Burnett, D. S., Casey, P., Everett, D. T., Lockhart, W., McComas, D. J., Mietz, D. E., MacNeal, P., Mireles, V., Moses, R. W. Jr., Neugebauer, M., Poths, J., Reisenfeld, D. B., Storms, S. A., and Urdiales, C.: 2003, 'The Genesis Solar Wind Concentrator'. *Space Sci. Rev.*, this volume.

Nishiizumi, K. and Caffee, M. W.: 2001, 'Beryllium-10 from the Sun', *Science* **294**, 352–354.

O'Mara, W. C., Herring, R. B., and Hunt, L. P. (eds): 1990, *Handbook of Semiconductor Silicon Technology*, Noyes Publications, Park Ridge N.J.

Powell, A., Petit, J.B., and Matus, L.G.,: 1991, 'Advances in silicon carbide chemical vapor deposition (CVD) for semiconductor device fabrication' NASA Tech. Memo 104410., prep for 1st Intl. High Temp. Elect. Conf., Albuquerque, NM (June 15–20, 1991)

Scott, C. E., Levinson, L. M., Maxwell, R. E., and Kaliszewski, M. S.; 1995, 'Solid state thermal conversion of polycrystalline alumina to sapphire' US Patent #5,451,553, 5–6.

Stansbery, E. K., Cyr, K. E., Allton, J. H., Schwarz, C. M., Warren, J. L., Schwandt, C. S., and Hittle, J. D.: 2001, 'Genesis Discovery Mission: Science canister processing at JSC', LPSC XXXII (cd-rom) 2064.pdf.
Treiman, A. H.; 1993, 'Curation of solar wind collector plates from a solar wind sample return (SWSR) spacecraft mission', JSC doc. #26406.
Wiens, R. C. and Burnett, D. S.: 1999, 'The Genesis Web Page', http://www.gps.caltech.edu/genesis.
Wiens, R. C., Neugebauer, M., Reisenfeld, D. B., Moses, R. W. Jr., and Nordholt, J. E.: 2003, 'Genesis Solar Wind Concentrator: Computer Simulations of Performance Under Solar Wind Conditions'. *Space Sci. Rev.*, this volume.

THE GENESIS SOLAR WIND CONCENTRATOR

JANE E. NORDHOLT[1], ROGER C. WIENS[1], RUDY A. ABEYTA[1],
JUAN R. BALDONADO[1], DONALD S. BURNETT[2], PATRICK CASEY[3],
DANIEL T. EVERETT[1], JOSEPH KROESCHE[4], WALTER L. LOCKHART[5],
PAUL MACNEAL[6], DAVID J. MCCOMAS[3], DONALD E. MIETZ[1],
RONALD W. MOSES, JR.[1], MARCIA NEUGEBAUER[6], JANE POTHS[1],
DANIEL B. REISENFELD[1], STEVEN A. STORMS[1] and CARLOS URDIALES[3]

[1]*Los Alamos National Laboratory, Los Alamos, NM, U.S.A.*
[2]*California Institute of Technology, Pasadena, CA, U.S.A.*
[3]*Southwest Research Institute, San Antonio, TX, U.S.A.*
[4]*Mobius Systems, San Antonio, TX, U.S.A.*
[5]*Creative Circuitry, San Antonio, TX, U.S.A.*
[6]*Jet Propulsion Laboratory, Padadena, CA, U.S.A.*
Author for correspondence, E-mail: JNordholt@LANL.gov

Received 15 December 2001; Accepted in final form 21 May 2002

Abstract. The primary goal of the Genesis Mission is to collect solar wind ions and, from their analysis, establish key isotopic ratios that will help constrain models of solar nebula formation and evolution. The ratios of primary interest include $^{17}O/^{16}O$ and $^{18}O/^{16}O$ to $\pm 0.1\%$, $^{15}N/^{14}N$ to $\pm 1\%$, and the Li, Be, and B elemental and isotopic abundances. The required accuracies in N and O ratios cannot be achieved without concentrating the solar wind and implanting it into low-background target materials that are returned to Earth for analysis. The Genesis Concentrator is designed to concentrate the heavy ion flux from the solar wind by an average factor of at least 20 and implant it into a target of ultra-pure, well-characterized materials. High-transparency grids held at high voltages are used near the aperture to reject >90% of the protons, avoiding damage to the target. Another set of grids and applied voltages are used to accelerate and focus the remaining ions to implant into the target. The design uses an energy-independent parabolic ion mirror to focus ions onto a 6.2 cm diameter target of materials selected to contain levels of O and other elements of interest established and documented to be below 10% of the levels expected from the concentrated solar wind. To optimize the concentration of the ions, voltages are constantly adjusted based on real-time solar wind speed and temperature measurements from the Genesis ion monitor. Construction of the Concentrator required new developments in ion optics; materials; and instrument testing and handling.

1. Introduction

The goal of the Genesis mission is to determine the elemental and, in particular, the isotopic composition of the original solar nebula. Because the Sun contains more than 99.5% of the solar system mass and because the outer layers of the Sun are thought to be relatively unchanged in composition over time, the solar photospheric composition is essentially equivalent to the composition of the solar

Space Science Reviews **105:** 561–599, 2003.

nebula. While some fractionation (preferential ion selection processes) undoubtedly occurs in the formation and acceleration of the solar wind (Bochsler, 2000; Bodmer and Bochsler, 2000), we believe this can be modeled and constrained by the collection of different solar wind flow regimes (Burnett *et al.*, 2003) to an accuracy sufficient to place significant limits on the solar nebula constituents and evolution. A definitive measure of this composition would serve as a baseline for understanding differences in the present composition of solar system bodies. Key isotopic ratios include the $^{17}O/^{16}O$, $^{18}O/^{16}O$, $^{15}N/^{14}N$, $^{13}C/^{12}C$, and noble gas isotopic ratios. Other objectives include the Li/Be/B elemental and isotopic abundances, which should help constrain models of solar evolution, and the radioactive nuclei (^{14}C, ^{10}Be), which should constrain models of the recent solar-surface energetic particle flux. Oxygen isotopic heterogeneity among solar system bodies (e.g., Clayton, 1993), cannot be interpreted without putting it in the context of the solar oxygen isotopic composition (Wiens *et al.*, 1999). Likewise, the evolution of planetary atmospheres, particularly in terms of the change in ratio of ^{15}N to ^{14}N and the ratios of the noble gas isotopes, cannot be correctly modeled without knowledge of the primordial composition (cf., Pepin, 1991).

In-situ mass spectrometric analysis of the solar wind isotopic composition is hampered by the large range of charge states and energies within the solar wind and especially by the large required dynamic range (e.g., $^{17}O/^{16}O \approx 4 \times 10^{-4}$). While lower precision measurements of N (Kallenbach *et al.*, 1998), O (Harris *et al.*, 1987; Collier *et al.*, 1998; Wimmer-Schweingruber *et al.*, 2001), Ne (Kallenbach *et al.*, 1997), Si, Fe, Cr, and a few other isotopes have been made by *in-situ* mass spectrometers, the Genesis mission takes a different approach by collecting samples of the solar wind to return to Earth for analysis.

The idea of implanting solar wind ions in target materials that are returned to Earth has a long and distinguished heritage dating back to the first published analysis of the methodology (Bühler *et al.*, 1966, 1969). As part of many of the Apollo landings on the Moon, thin foils of Al or Pt were exposed to the solar wind on the lunar surface and returned to Earth (cf., Geiss *et al.*, 1969, 1970, 1972). These measurements are today still benchmarks of the noble gas isotopic composition of the solar wind. The technique of increasing solar wind fluence with electrostatic fields was also discussed by Signer (1965). The Concentrator builds on this concept to produce an electrostatically focused solar wind collector that uses ultra-pure targets for sample collection and return. The basic concentrator concept, as originally proposed for Genesis, is described by McComas *et al.* (1998).

The Genesis mission was launched in August 2001 and traveled to the L1 point between Earth and the Sun where it is exposing collector materials to the solar wind, a mission segment that will last over 2 years, and will then return the samples to Earth for ground-based analysis. After sample return in fall 2004, analyses of target materials will take place using a number of techniques. The isotopic abundances of the implanted solar wind in these samples and by proxy the solar

TABLE I

Target implantation levels expected in the Concentrator. Concentrations refer to average over outer 100 nm for a 26 month exposure. Abundance fractions from Bochsler and Geiss (1989).

Element	Solar wind abundance, atomic fraction	Expected concentration given 20× concentration factor
N	8×10^{-5}	4×10^{18}/cc
O	5×10^{-4}	3×10^{19}/cc
F	~ 20 ppb	1×10^{15}/cc
Ne	9×10^{-5}	4×10^{18}/cc

composition will be determined to levels of accuracy required by planetary science applications.

The primary difficulty with the Genesis concept is the low fluence of ions heavier than helium in the solar wind. Measuring the isotopic abundances of some key elements such as O and N is difficult using simple passive collection. This is because large (>50 cm^2 areas) must be analyzed and because only one material (float zone Si) has adequate bulk purity. However, float zone Si has a native oxide layer and, because of SiN formation, extraction N from large areas of Si is difficult. Consequently, to optimize signal to background, allowing the determination of these isotopic ratios, the Genesis payload includes a large-aperture solar wind Concentrator designed to increase by an average factor of ~ 20 the fluence of isotopes in the m/q range of 2.0 to 3.6 amu/charge (essentially masses 4 to 28). The Concentrator also increases the average energy of ions entering the target, implanting solar wind ions below the native oxide layer. Table I lists estimated elemental abundances normally seen in the solar wind and their expected implantation levels in the Concentrator targets.

2. Overview

The Genesis spacecraft is designed to orbit the L1 point between the Earth and the Sun for nearly 28 months while collecting solar wind. It is pointed into the solar wind $4.5 \pm 2.0°$ off a direct line of sight to the Sun such that it will be directly facing the apparent average solar wind direction. The spacecraft will then return to Earth and its sample return capsule will parachute back to Earth where the samples that have been implanted with solar wind will be removed and distributed to laboratories for elemental and isotopic analysis.

Figure 1. The Concentrator in its flight configuration is visible inside the science canister. During the spacecraft's journey to and return from the L1 point and during the re-entry into Earth's atmosphere the collector arrays are stowed over the Concentrator and the lid containing the bulk collector array is closed forming an airtight seal to protect the exposed samples from contamination.

The Genesis Concentrator is a unique instrument in that it does not send back any data from space other than simple state-of-health housekeeping data. Instead, its targets will be removed for analysis at the end of the mission. Figure 1 shows the Concentrator inside the spacecraft's science canister which prevents contamination during the launch, flight, and re-entry phases of the mission. Once on station at L1 the science canister is opened as shown to expose the Concentrator to the solar wind.

The Concentrator includes many novel developments: a new design for electrostatic mirrors; new testing methodologies; ultra-clean shipping and handling in the final stage of flight preparation; ultra-fine, ultra-strong stainless steel mesh; O-free resistive coatings for ceramic components; and high precision segments of ultra pure SiC, ^{13}C enhanced CVD diamond, and low N amorphous diamond film on a Si substrate (Jurewicz *et al.*, 2003).

2.1. Operation

Figures 2 and 3 show the Concentrator's final design in partially cut-away cross-section and cut-away 3-dimensional views. Ions from the solar wind come from the upper portion of these figures, the direction pointed into the solar wind in the operational configuration. Ions initially traverse the first, grounded grid to enter the Concentrator. The outermost grid and instrument skin must be at spacecraft

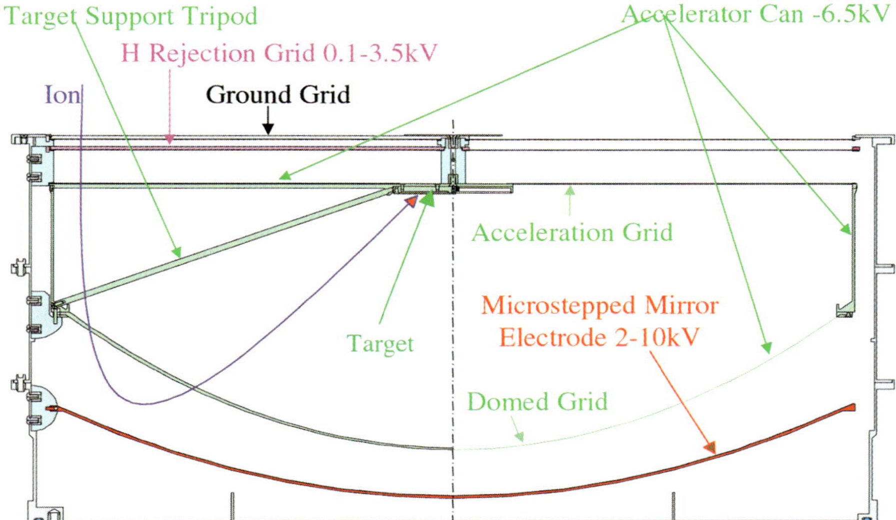

Figure 2. Cross-section of the Concentrator showing the major elements of its design. Electro-optic elements that share voltage levels are color coded to indicate which parts are electrically connected. The left-hand side shows a cross-section through the grid supports and insulators, while the right-hand side shows the cross section between insulators and grid support ribs.

ground to prevent stray electric fields from the Concentrator from interfering with other parts of the spacecraft. The next grid down is the hydrogen-rejection grid. This grid is biased such that at least 90% of the solar wind protons are rejected in order to minimize proton radiation damage to the target. Rejection of this fraction of solar wind protons without also rejecting heavy ions requires that the hydrogen rejection grid voltage maintains a constant ratio relative to the peak of the solar wind proton energy distribution. The Genesis ion monitor (Barraclough *et al.*, 2003), together with the science algorithm (Neugebauer *et al.*, 2003) provide the energy levels every 2.5 min to update rejection grid voltages. The upper two panels in Figure 4 show how variations in the voltage on the hydrogen-rejection grid change the amounts of hydrogen and heavy ions rejected vs. electrostatic ratio of voltage to the peak of the solar wind proton energy distribution. The voltage on the hydrogen-rejection grid is $V_H = Ep^*R$ where Ep is a running average of three measurements of the peak of the proton energy distribution from the ion monitor (Barraclough *et al.*, 2003), and R is the ratio. The lower panel of Figure 4 shows the fractionation, or change in the isotopic ratios due to the Concentrator, of oxygen as a function of ratio R and thermal Mach number. The losses and fractionation limit the voltage level at which the hydrogen-rejection grid can be operated and still achieve the desired collection of heavy ions. A thorough analysis of the mass fractionation in the Concentrator is presented in Wiens *et al.* (2003).

Ions which successfully transit the hydrogen-rejection grid, will next have their group velocity, the velocity of energy propagation, increased relative to its value in

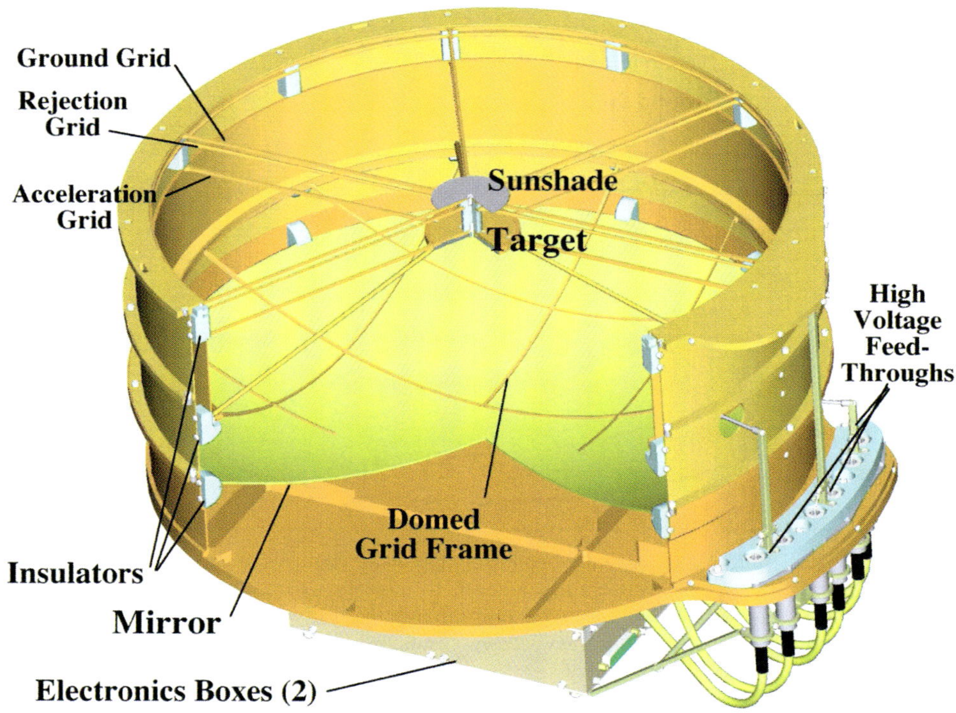

Figure 3. Three dimensional cut-away view of the Concentrator showing its major mechanical and electro-optical components (for clarity only the grid supports and not the grids themselves are shown). The Concentrator electronics boxes and the cabling leading from them to the hermetically sealed high-voltage feed-throughs and the high voltage input section where voltage is distributed to the Concentrator's electro-optical elements are clearly visible. The high voltage input section is shown without its cover which is attached to the outer skin of the Concentrator for flight.

the solar wind while their relative angular scattering is decreased as they approach the 'acceleration' grid. The acceleration energy acquired by ions as they approach the acceleration grid is $E_a = -qV_a$ where q is the ion's charge and V_a is the voltage applied to the acceleration grid. The total energy of an ion is $E_t = -qV_a + E_i$ after it passes through the acceleration grid, where E_i is the ion's original energy in the solar wind. The acceleration accomplishes two goals: (1) it straightens the ion trajectories, allowing higher concentration factors, and (2) it provides higher energy implantation into the target. The acceleration voltage will be fixed at -6.5 kV during operation and -10 kV is the maximum allowed in the design. After passing through the acceleration grid, ions enter the field-free acceleration 'can', a region in which all ions maintain their trajectories and energies of $-qV_a + E_i$. Having passed through the acceleration can, the ion must then pass through the 'domed' or mirror grid which is part of the acceleration can and is maintained at the same voltage as the acceleration grid. The acceleration can is clearly visible in Figures 2 and 3 and is made up of the acceleration grid, the domed grid, and the outer can

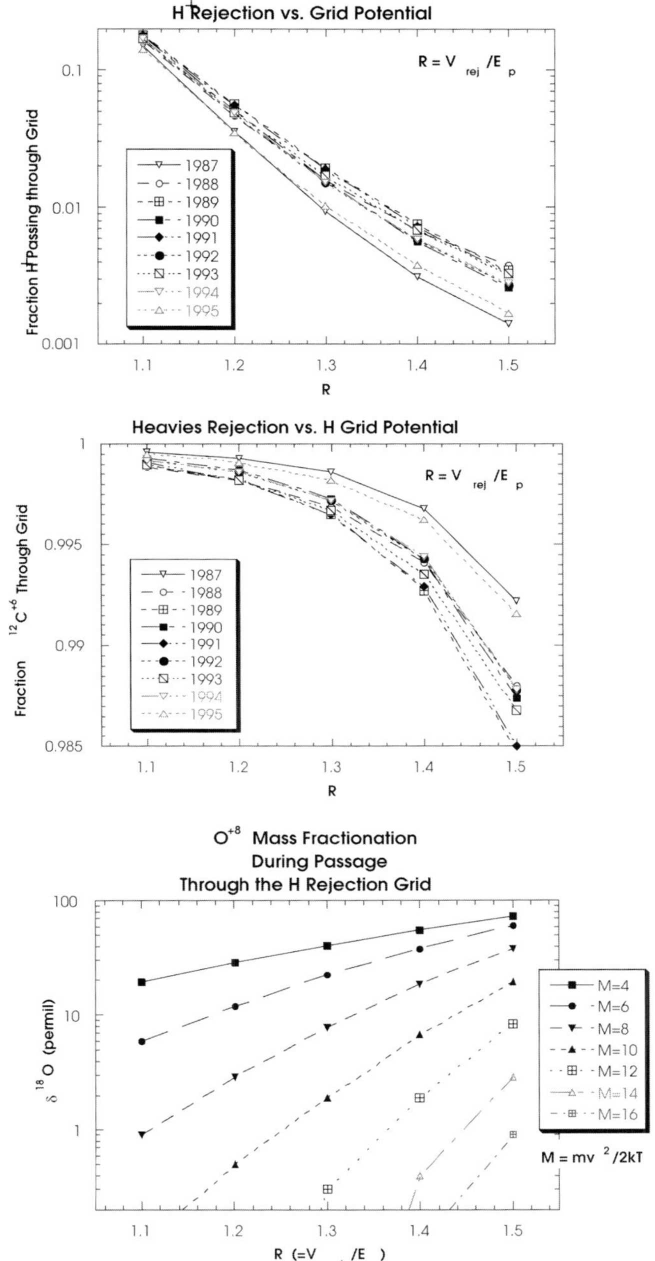

Figure 4. The upper panel shows the fraction of H^+ rejected by the hydrogen-rejection grid for a given ratio, R. The middle panel shows the expected loss of $^{12}C^{+6}$ ions vs. R. With an m/q of 2.0 amu/charge, $^{12}C^{+6}$ is indicative of the worst-case losses induced by the hydrogen-rejection grid. C itself cannot be measured in the Concentrator targets which all have C as a component. The lower panel shows the mass fractionation of O introduced by the hydrogen-rejection grid as a function of ratio and thermal Mach number. These panels give the calculated effects of the hydrogen-rejection grid for average solar wind conditions for the years 1987-1995.

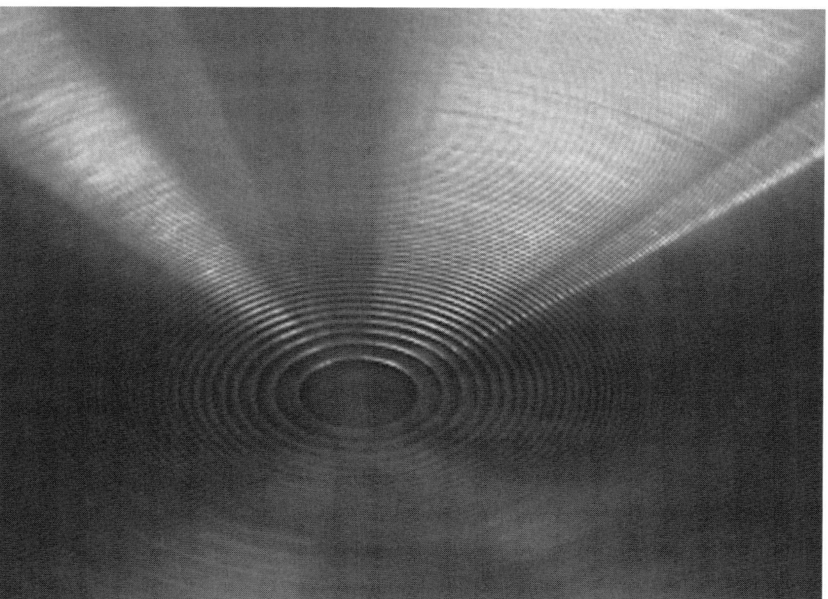

Figure 5. The Concentrator's mirror electrode which can be operated at up to +10kV to produce, in conjunction with the domed mirror grid, a reflecting field which focuses ions onto the Concentrator target. The bottom panel shows an enlarged view of the mirror's center where the microstepping, which prevents this electrode from focusing light onto the target, can be seen most easily.

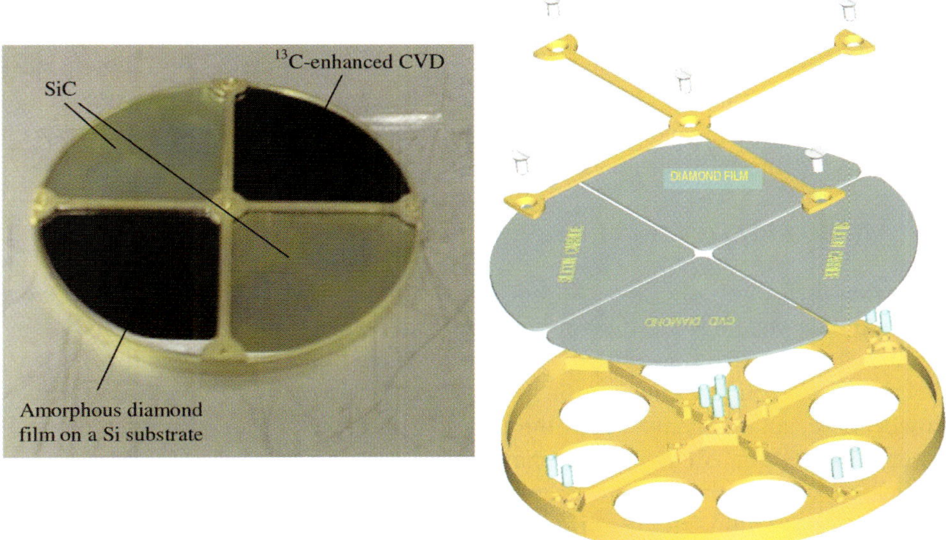

Figure 6. The Concentrator target is shown fully assembled at left and in an exploded engineering drawing at right. The SiC, CVD diamond, and amorphous diamond film on Si sections are clearly visible. The drawing on the right labels the sectors as they are arranged in the flight configuration. Also visible in blue in the drawing at right are the springs which apply force to the corners of the target sectors holding them firmly during vibration.

wall shown in green in Figure 2. Ions are reflected in the region between the domed grid and the bottom 'mirror' electrode which is maintained at a positive voltage of up to +10 kV. The region between the mirror grid and the mirror electrode forms a 'mirror field' which acts to reflect ions and focus them onto the target. Like the hydrogen rejection grid, the mirror electrode voltages are adjusted in real time based on monitor and science algorithm data. The voltage on the mirror electrode is set to give a 20% margin in energy on the highest expected oxygen m/q, of 3.6, based on $^{18}O^{+5}$, therefore the mirror voltage is $V_m = 4.32^* Ep$ and will produce a mirror point in most ion's trajectories well away from the mirror electrode. Ion trajectory turn-around far from the mirror is necessary because the mirror electrode itself is not a smooth surface. It has been microstepped as shown in Figure 5 so that optically it appears to be a flat plate reflecting the Sun's photons back to space but the electric field it produces at a distance several times the height of the microsteps (100 μm) will smoothly focus the solar wind ions. Microstepping is a necessity if the target temperature is to be below $\sim 250\,^{\circ}$C to prevent implanted ions from being boiled off. Testing showed that less than one percent of the total incident optical flux was focused onto the target region by the mirror electrode. Ions reflected by the mirror field travel back into the acceleration can transiting the domed grid once again. Having re-entered the acceleration can the ions again have the total energy

E_t described above. Ions are now focused onto the target area in which they are then implanted at a depth which has been increased by the additional acceleration component, $-qV_a$, of their total energy. The target's sensitive area, which faces downward in Figures 2 and 3, is made up of SiC, amorphous diamond film on a Si substrate, and ^{13}C enhanced CVD diamond sections (Burnett *et al.*, 2003) which are clearly visible in Figure 6.

2.2. Mechanical/cleanliness requirements

The grids that carry the Concentrator's operating voltages and form the desired electric fields must have as high an open area ratio as possible while still maintaining structural integrity and highly uniform electric fields. A grid pitch of at least 20 lines cm^{-1} (50 lines inch^{-1} or lpi) is necessary to maintain field uniformity. However, because ions must pass through 5 grids (twice through the domed grid) to reach the target, all grids must have open area ratios $A \geq 0.90$ because the percentage of ions that transit all grids is geometrically A^5. Thus the grid wires can have a maximum thickness in the direction parallel to the surface of the grid of ≤ 26 μm. The grids used in the Concentrator are woven from 25 μm stainless steel wires on a 50 lpi pitch which gives them an open area ratio 90.4%. All structures supporting the grids and elsewhere in the Concentrator flight region were likewise minimized to prevent the loss of ions from collisions with these structures. Although grid support frames are only a few millimeters thick these structures also can perturb the desired electric fields and were placed on the side of the grids that minimized the field variations they cause. For the grounded grid the structure is above the grid in the low field region around the spacecraft. The hydrogen-rejection grid has, on average, a lower field strength on the side closer to the grounded grid, therefore the structure supporting the hydrogen-rejection grid is on the side nearest to the grounded grid. The volume inside the acceleration can is a field-free region therefore the acceleration grid's supporting structure is below its grid and the support for the domed grid is above its grid.

The ultra-clean requirements mean that the electronics must be sealed away from the Concentrator and the internal portion of the sample-return canister. The electronics boxes were therefore mounted on the bottom side of the Concentrator's base plate. The base plate forms a seal with the sample-return canister's deck using a low vapor pressure o-ring. All electronics connections are made through hermetic feed-throughs welded into the Concentrator's base plate and all of the feed-through parts and wiring for the high voltages on the clean side of the Concentrator base plate are ultra-low outgassing ceramic or metal components. By using the hermetic seal, standard flight electronics could be used without causing contamination to the Concentrator or any other collection materials mounted inside the sample-return canister. Voltages are input to each electro-optical component on hard connections which are wired to each of the feed-throughs with soldered connections to small wire jumpers. The solder was allowed despite cleanliness requirements because a low outgassing solder was used and, as is visible in Figure 3, the high voltage input section is largely sealed off from the rest of the Concentrator. Connections between the high voltage bus bars, small wire jumpers, and high voltage feed-throughs can be seen in Figure 3.

2.3. Testing

The Concentrator was subjected to all of the standard tests required of a flight instrument as well as being subjected to a unique characterization regimen designed to provide flight performance verification. After initial tests of a prototype confirmed that the Concentrator design would meet the scientific requirements, two Engineering Models (EMs) were constructed. Two EMs were required to meet the demanding integration and test schedule. One model was used for detailed thermal testing, including solar thermal vacuum grid mapping of the Concentrator alone (instrument level), and solar thermal vacuum testing of the Concentrator while integrated into the spacecraft canister as shown in Figure 1 (canister level). The use of two EMs allowed modification of the thermal EM for installation of thermocouples and associated wiring into the interior portions, including the target and grid frames. The flight model (FM) was tested at instrument and canister levels prior to final cleaning and gold coating, though as such it had to forego accurate thermal testing. This was acceptable because of the high fidelity of the EMs. After re-assembly, the FM was never exposed outside a clean room or clean vacuum chamber. This necessitated the unusual omission of a final workmanship vibration, but all of the final assembly techniques had been tested and verified several times on the EMs and were both performed and observed by multiple quality assurance personnel. Its flight-readiness was verified by high voltage testing to 105% of maximum operating voltages and a thorough final optical mapping which can be done under clean conditions and is described below.

3. Design and Development

The design goals of a concentration factor of at least 20 with a mass fractionation averaged across the target quantified to better than 0.1% was set to allow sufficiently detailed analysis of implanted material to provide the solar isotopic ratios of the $^{17}O/^{16}O$ and $^{18}O/^{16}O$ to $\pm 0.1\%$, and $^{15}N/^{14}N$ to $\pm 1\%$ using the best analytic techniques available. These design requirements were primarily developed by comparing what could be collected from the solar wind in a reasonable mission length with the minimum content of O and N in ultra-pure materials and determining what would be required to analyze implanted O and N accurately. Besides concentration, the Concentrator must not introduce unknown levels of mass fractionation or fractionation gradients in the target that would inhibit the accurate analysis of the target. Table II shows the full list of performance criteria that were

TABLE II

Concentrator performance requirements.

Performance parameter	Value required	Value achieved	Comments
Average concentration Factor for N and O	≥ 20	22.8	Low speed solar wind
	20.9		Weighted average over all expected solar wind speeds
Target area	≥ 15 cm^2	25.9 cm^2	
Errors in $^{17}O/^{16}O$	0.1%	0.085%	Calculated with SIMION before flight to this level Checked after flight with Ne^+ implantation levels
Target temperature	≤ 250 °C.	230 °C	Worst case calculated temperature; EM STV test target frame = 144 °C
Ion acceleration before impacting target	> 5 kV	10 kV possible see text	Operation at 6.5 kV
Solar wind proton fluence prevented from reaching target	$\geq 90\%$	$\geq 93\%$	Averaged over all thermal Mach numbers
Surface contamination by C, N, O	$<10^{15}$ atoms cm^{-2} under ultra high vacuum ($<10^{-8}$ torr) conditions at 200 °C		Concentrator gold coating approved. Grids not coated – calculations show minimal implantation resulting. Inflight contaminates do not implant.

used throughout the development of the Concentrator. A numerical limit was not placed on the fractionation gradient but it has been checked to be small enough that it does not cause target samples to be analyzed in unworkably small radial segments (Wiens, 2003).

3.1. Mirror Design

Many different electro-optical configurations were tested but the most efficient is the ion analog of the optical reflecting telescope described above. This is because like an optical telescope, an ion telescope concentrates the flux which impinges on its aperture by using a large collection area well-focused onto a relatively small target. Designs which use multiple targets and apertures or electro-optical lenses to focus ions onto target areas were examined but ultimately the best concentration is achieved just as it is in telescopes – a large-area aperture, most compactly and

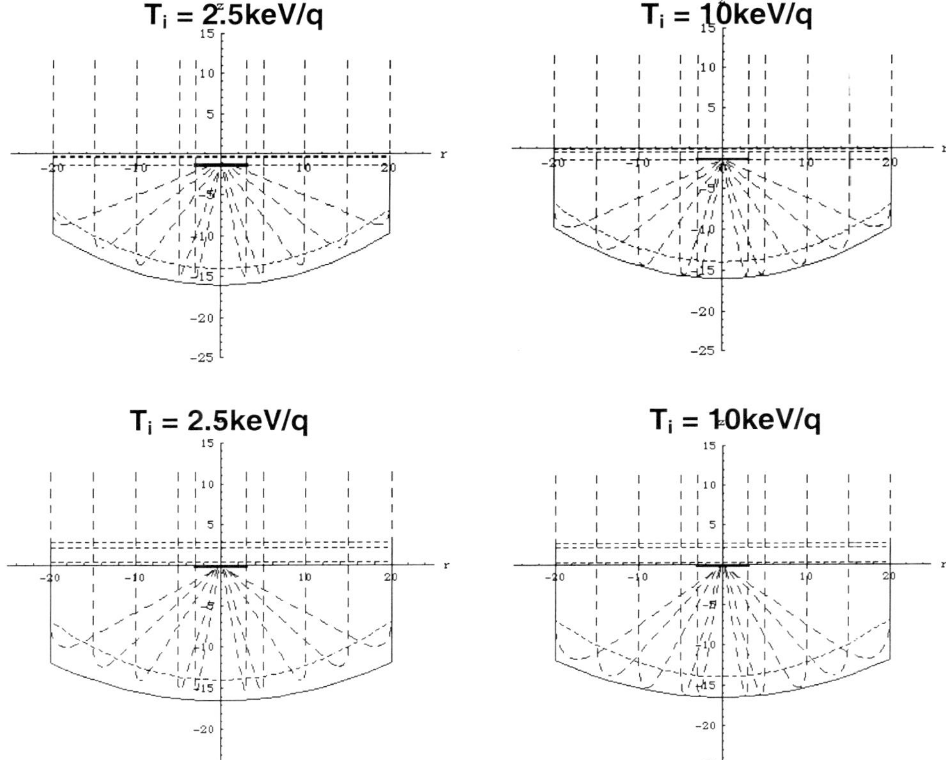

Figure 7. Comparison of Concentrator electro-optical designs for ions of different energies. The design in the upper two panels uses confocal parabolas for the domed grid and mirror electrode, but it is energy-dependent. Although it focuses ions with incident energy of T_i = 10keV/q it focuses poorly at 2.5keV/q. The Concentrator's optimized design which maintains its focus over all input energies is shown in the lower two panels. This design drastically reduces the expected isotopic fractionation.

cheaply produced with a mirrored configuration, is focused with low f-number optics onto a small target. The 6.2 cm diameter of the Concentrator's target comes directly from the focusing ability of the ion mirror given the known angular and energy distributions of the solar wind. However, unlike with photon-optics, ion-optical mirrors that give low f-number focusing cannot be made achromatic with simple parabolic optics. The equivalent of photon-chromatic effects in ion optics is differential focusing which is energy-dependent. This is a major concern for the Concentrator because the solar wind accelerates all of its components to the same velocity and this means that any mass difference in component ions translates to an energy difference. Energy-dependent focusing is thus equivalent to mass fractionation which is unacceptable for the scientific investigation to be performed by the Concentrator.

TABLE III

Some isotopes, charge-states, and m/q ratios found in the solar wind.

Isotope	Charge-state	m/q
H	1	1.0
^{4}He	2	2.0
^{16}O	8	2.0
^{18}O	8	2.25
^{18}O	5	3.60
^{22}Ne	8	2.75
^{56}Fe	9	6.22

This was corrected by developing an electro-optical design which produced a virtual parabolic surface at which the ions were reflected. The upper portion of Figure 7 shows the energy-dependent or chromatic focus of a design where similar parabolic mirror grid and electrode designs are used. No design of this type with similar parabolic electro-optical mirror elements will focus without chromatic distortions because the penetration of the ions into the mirror-field region varies with the angle between the mirror surfaces and the incident ion's direction. Ions of equal energy will penetrate less deeply into the mirror region at the edges of the mirror because the reflection point is determined by the component of particle's velocity which is parallel to the field lines. Thus ions in the more steeply inclined portion of the mirror near the edges turn around closer to the mirror grid than those in the central portion. As Figure 7 shows, this effect varies with ion energy and thus cannot be corrected by a simple adjustment of the focal length of the parabola. The lower half of Figure 7 shows the achromatic design developed for the Concentrator. In this design the loci of reflection points is parabolic and the surfaces of the mirror grid and mirror electrode diverge toward the edges of the Concentrator. This reduces the electric field in this region and compensates for the reduced velocity component parallel to the field lines.

3.2. The Hydrogen-Rejection and Acceleration Grids

While energy-dependent focusing has been eliminated from the ion mirror, there are still energy dependent effects associated with the hydrogen-rejection grid and the acceleration can. Since the voltages of both of these electro-optical components operate directly on only one component of an ion's velocity vector, they alter the trajectories of incoming ions. This energy-dependent effect cannot be corrected by adjusting the electric fields around the grids in the way that is done for the mirror, and constitutes the bulk of the remaining mass fractionation in the Concentrator.

TABLE IV

The voltage setting on the hydrogen-rejection grid will be determined from the peak of the proton energy distribution and the thermal Mach number determined from the solar wind data provided by the Genesis Ion Monitor (Barraclough *et al.*, 2003).

M	R
> 11	1.3
$11 > M > 9$	1.2
$9 \leq M \leq 8$	1.1
$8 > M > 6$	1.0
≥ 6	0

H rejection voltage $V_H = Ep^*R$
$M^2 = mv^2/2kT$.
Ep is the peak in the proton energy distribution averaged over the last three consecutive measurement cycles.

This problem is further compounded by the multiple charge states present in the solar wind and the fact that forces on the ion are proportional to an ion's charge multiplied by the grid voltage. Table III gives a sampling of the charge states for some isotopes found in the solar wind and their mass per charge ratios.

Figure 8 gives a schematic overview of the effects of the acceleration grid on the ion distribution hitting the target. The effect of the acceleration voltage is to narrow the angular distribution of ions with lower m/q more than the angular distributions of ions with higher m/q. This causes ions with lower m/q to be implanted more centrally in the target than ions with higher m/q. The hydrogen rejection grid acts to widen the angular distribution for ions with lower m/q more than for ions of higher m/q. The effects of the angular distribution changes introduced by the hydrogen rejection and acceleration voltages are not mutually cancelled out. Figure 9 shows the effect of different acceleration voltages on mass fractionation across the target. The planned operational acceleration voltage has been reduced from -10 kV in the original flight plan to -6.5 kV to reduce this effect. The final operational setting for the acceleration voltage was the result of a trade study in which the advantage of a lower mass fractionation gradient across the target is traded against a slightly higher concentration factor, lowered ion backscattering losses at the target surface, and deeper implantation depths. Mass fractionation resulting from the hydrogen-rejection grid is less easily reduced. Figure 4 shows the levels at which its voltage must be operated to reject the required 90% of the incoming hydrogen. This criterion cannot be relaxed without potentially damaging the target. The added mass fractionation that different levels of hydrogen-rejection

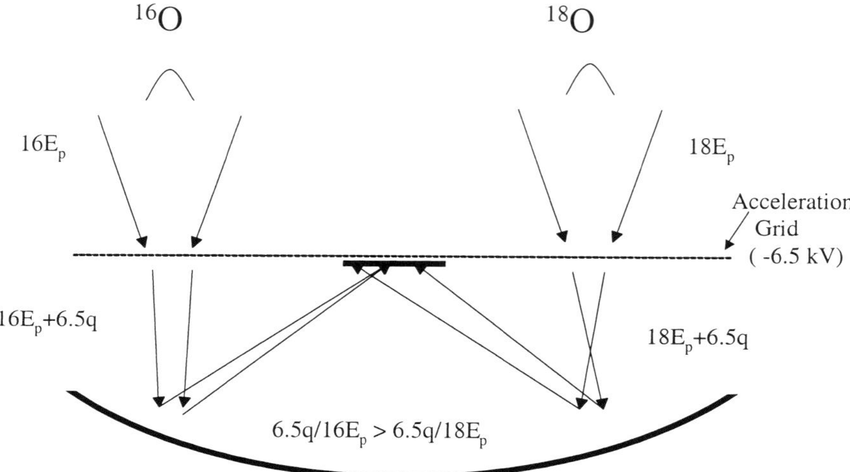

Figure 8. An exaggerated drawing illustrating the different effects of the acceleration grid on ^{16}O and ^{18}O. External to the Concentrator the isotopes have the same angular distribution, but the acceleration grid is more effective at straightening the distribution of ^{16}O. ^{16}O is then more tightly focused than ^{18}O, leading to a fractionation gradient on the target, with more ^{16}O concentrated at the center.

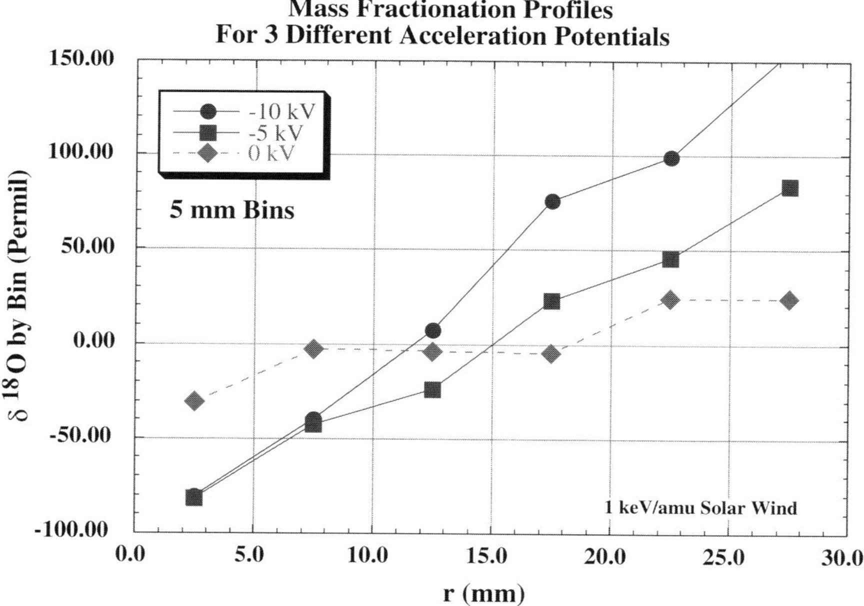

Figure 9. The permil mass fractionation of ^{18}O at a given radius across the target as produced by different operating voltages on the acceleration grid. The simulation used a perfect parabola rather than the actual mirror grid shape, and as such it is a worst case illustration of the effect of the accleration grid (cf. Wiens *et al.*, 2003).

TABLE V

Temperatures measured at different parts of the Concentrator during solar thermal vacuum testing.

Component	JPL EM test °C
Target shade	93
Target frame	144
Mirror	132
Accel. can	105
Accel. frame, in sunlight	122
Accel. frame, out of sunlight	105

grid voltages produce is discussed in detail in a companion paper to this one (Wiens *et al.*, 2003). Table IV gives the parameters that will be used to set the voltage on the hydrogen rejection grid.

3.3. Operating Temperature

After the electro-optical design was complete, the Concentrator development team had to determine what temperature control measures were needed and how best to implement them in an ultra-clean instrument. The requirement that the Concentrator not introduce additional oxygen or nitrogen into the target from sputtered ions meant that the instrument's internal surfaces had to be coated with a material that would not form a native oxide layer or introduce N-bearing compounds. Gold is the most easily obtainable coating that meets these criteria. Unfortunately gold has a high thermal absorption to emission (alpha/epsilon) ratio, which causes it to heat to an unacceptably high temperature in some parts of the Concentrator. A careful examination of other materials which are N-free and do not form native oxide layers established that they all had similar alpha to epsilon ratios. Active cooling could not be provided without violating the cleanliness requirements of the Concentrator (e.g., a micro-meteoroid-induced pinhole in a heat pipe would introduce contamination from coolant leakage) so other temperature mitigation methods had to be applied. The components of the Concentrator which have critical temperature control requirements are the target, the Concentrator Electronics Boxes (CEBs), and the grids. The most important element for which the operating temperature must be controlled is the target. The maximum target temperature above which diffusion would likely cause loss of the solar wind sample is 250 °C. Microstepping the mirror electrode and placing the target behind an anodized sunshade (anodized materials have low alpha/epsilon ratios) reduced its operating temperature to an acceptable 230 °C, as a worst case calculation. The anodized coating on the sunshade is acceptable because it is external to the Concentrator's focusing elements and their structures. The CEBs are mounted on the anti-solar side of the sample-return canister and, with sufficient thermal control of the Concentrator's mounting plate, can be kept below an acceptable maximum operating temperature of 77 °C. Temperatures measured in the solar thermal tests performed on one of the Concentrator EMs are given in Table V.

Thermal measurements and control of the grids and their supports presented several different challenges. The mirror electrode and grid support frames were mounted in a self-centering fashion on a series of pins in radial grooves that allowed them to expand relative to the Concentrator's outer can which is cooled by radiation and by connection to the Concentrator's baseplate. The baseplate is also in contact with the temperature controlled sample-return canister deck. The grids were produced from the same type of stainless steel as the frames but the inability of the tiny wires that make up the grids ($\varphi = 25\ \mu$m) to efficiently radiate their heat to space in comparison to the larger frame members means that the grids are substantially warmer when in sunlight. This raises concerns about grid wrinkling and deflection by the electrostatic forces on them from their high voltages. Further, when the spacecraft is turned so that the Concentrator moves from sunlight to shadow, the fine grid wires, lacking thermal mass, cool immediately while the frames do not. This means the grids wrinkle in the sun and are severely stressed when moved into shadow. Grids need to be strong and very firmly attached to their frames without forcing the frame size to be large thereby blocking ions. Preliminary tests of grids mounted on hoops showed that woven mesh was far stronger than etched mesh. The Concentrator prototype used hexagonal 24 lines cm^{-1} (60 lpi) etched mesh for the flat grids and woven for the domed grid but all later Concentrators (both EMs and the FM) used woven mesh for all grids. Attachment of the grid to their frames was shown to be sufficiently strong both for grids spot welded to their frames with a cover-shim or for grids clamped to their frames in concentric hoop mounts. For grids in areas critical for ion blockage, the hydrogen-rejection and acceleration grids, grids were spot welded to minimize frame size. Grids that had less critical areal constraints and were subject to more risks from handling, the ground and acceleration grids, were hoop mounted. Thus the domed grid and grounded grid were both hoop mounted while the acceleration and hydrogen-rejection grids were spot welded. In both cases grids were bonded to only the inner and outer rings, and not to the radial ribs, to allow for greater flexibility of the mesh during heating or cooling events. Tests with grids in a solar thermal vacuum chamber demonstrated that rigidly mounted woven mesh could withstand the stresses induced by the thermal shock which occurs when the Concentrator is moved from full sunlight to shade, however the high operating temperature of the grids in comparison to that of their frames and the consequent wrinkling and sagging of the grids was unacceptable if the grids were gold coated. Careful analysis of the likelihood of material sputtering off uncoated grids and implanting in the target showed that, largely because of the very high open area of the grids, they could be left uncoated. The increased thermal

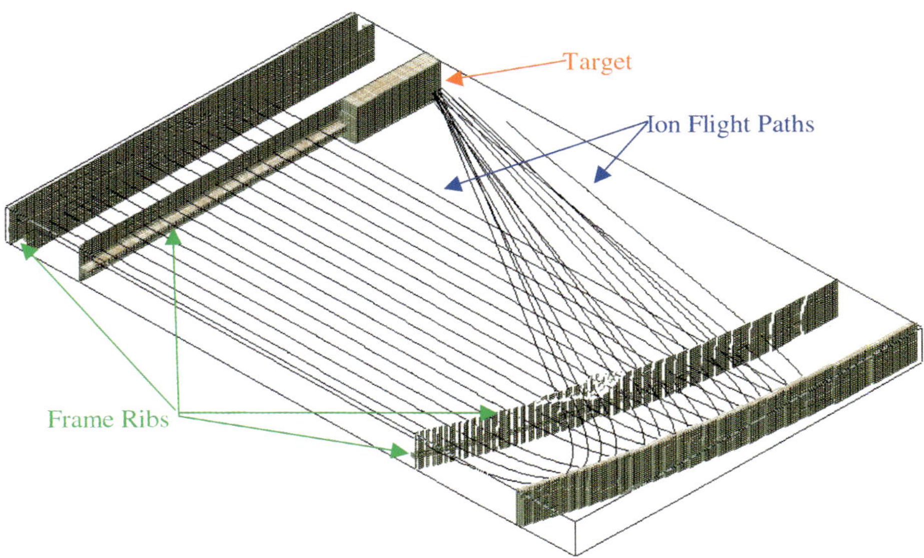

Figure 10. A cross sectional view of the Concentrator as it is modeled in SIMION. Ribs supporting each of the four grids can be seen along with the paths of several ions showing their reflection in the mirror region and implantation into the target.

emission of bare stainless steel relative to gold surfaces keeps the grids at a much lower temperature and does not cause significant wrinkling or sagging when in full sunlight.

3.4. DOMED GRID

Another major concern was the structural integrity of the domed grid. The waffle-patterned support shown in Figure 3 was found to be optimal for strength with minimal ion blockage. However when grid was mounted on this frame using a mandrel to produce a grid surface with the correct curvature, the grid was able to move and change its figure. There were concerns about how best to test the Concentrator mimicking the zero gravity and electrostatic forces on them. SIMION modeling showed that some defocusing caused by stretching the grid tightly over its frame, in fact, reduces mass fractionation and the fractionation gradient in the Concentrator target (Wiens *et al.*, 2003). The second engineering model and the flight model used grids stretched tightly over the domed grid frame. This configuration has very good structural stability under all of the conditions in which the Concentrator must operate. As shown in Figure 10, support structures and grids were included in SIMION models. Also an analytic function for scattering by the grids was developed and integrated into the SIMION models showing that the 20 lines cm^{-1} (50 lpi) grids and their supports did not unduly change the ion focusing of the Concentrator.

Figure 11. The EM Concentrator is installed in the ion beam test chamber. The central MCP assembly and cable harness can be seen inside a metal cover used to prevent stray fields from this assembly from causing deviations to incoming ion trajectories.

3.5. Flat Grids

The electrostatic forces on the grids were also examined. Although this is not an issue for the tightly stretched domed grid, the flat grids have long unsupported spans that help them to withstand the sudden temperature changes that occur when the spacecraft moves off sun pointing or covers the Concentrator with the collection arrays. The electrostatic forces on these grids are between 0.005 and 0.031 N (away from the mirror electrode for the acceleration grid and toward the mirror electrode for the hydrogen-rejection and grounded grids), while gravity on Earth exerts ~ 0.018 N on the grids. Functional testing of the Concentrator in a vertical position such as that shown in Figure 11 where the gravitational forces are orthogonal to the electrostatic forces indicated that the grids could maintain their shapes adequately when these forces are applied.

TABLE VI
Concentrator voltages.

Electro-optic component	Typical operating voltage	Maximum CEB voltage	EM test voltage
Hydrogen-rejection grid	1.0 kV (variable, 2.5 min time scales)	3.5 kV	4.55 kV
Acceleration can	−6.5 kV (fixed in flight)	−10 kV	−13 kV
Mirror electrode	4 kV (variable, 2.5 min timescales)	10 kV	13 kV

3.6. High voltage power supplies

The programmable voltages that run the Concentrator are produced by two Concentrator Electronics Boxes. Each CEB produces all 3 voltages needed to run the Concentrator (hydrogen rejection, accelerator, and mirror electrode voltages) and each voltage is individually controllable. Table VI lists all of the voltages needed in the Concentrator. Two CEBs are used to provide operational redundancy, which is further enhanced by separately wiring each CEB output voltage into the Concentrator. Each voltage is output onto a coaxial cable and input to the Concentrator through hermetic feed-throughs welded into the Concentrator base plate. The Concentrator is normally powered by CEB-1. In the event of any voltage failure, that CEB will be turned off and all voltages supplied by CEB-2. Because of the commanding scheme used, the two CEBs cannot be on at the same time but this is not prohibited by the high-voltage wiring configuration. Figure 3 shows the cabling, feed-throughs (produced by Reynolds Inc.), and high voltage input section. The jumper wires that join the connections from each CEB to the bus bar that distributes the voltage to the related electrostatic element are also clearly visible. The currents drawn are very modest, coming from two sources: (1) Photoelectrons produced at the ground and acceleration grids each contribute approximately −500 nA to the H rejection grid, and the mirror grid produces about −360 nA which is collected on the mirror electrode. (2) Resistive coatings on the insulators that hold the mirror electrode in place and support the target below the sunshade. These coatings, which are described more completely below, act as high value resistors connected to the mirror electrode, hydrogen rejection grid, and accelerator grid, drawing approximately 0.25 μA, 0.5 nA, and 0.5 nA, respectively. The CEBs are able to supply approximately 3 μA at each voltage, well above requirements. The CEBs underwent thermal vacuum tests at both the EM and FM levels by themselves to avoid contaminating the sensor head.

4. Testing and Characterization

The Concentrator has a unique design with a wide variety of requirements (see Table II) which are of primary importance to the scientific goals of the Genesis mission. As such 4 Concentrators (1 prototype, 2 EMs, and 1 FM) were built and subjected to a wide variety of tests. In addition to standard vibration, thermal vacuum (instrument mounted in vacuum with temperature variations on the mounting surfaces to determine overall operational temperature range), solar thermal vacuum (instrument mounted in vacuum and exposed to a solar optical source to determine operating conditions in space), and thermal cycling, numerous tests were developed specifically for the Concentrator.

The Concentrator is designed to collect solar wind ions using voltage levels calculated from solar wind speed derived from the onboard solar wind ion monitor data (Neugebauer *et al.*, 2003; Barraclough *et al.*, 2003). Its actual data will be collected by analyzing ions implanted in its target well after the mission is completed. Only a very limited number of flight-like targets could be analyzed to test the Concentrator's operation. Therefore in order to fully test and characterize the Concentrator it was necessary to develop several new methods of analyzing the implantation efficiency into the Concentrator's target and the overall operability of a Concentrator. For these purposes 4 methods of testing and verifying Concentrator operation were developed: micro-channel plate detector (MCP) testing; foil implantation; grid and surface mapping; and target implantation.

4.1. MCP TESTING

The first method, shown in Figures 11 and 12, replaces the target with a set of MCPs and was developed to allow comparison of simulated performance with actual performance. The upper 3 grids of the Concentrator must be replaced with grids on frames specially designed to hold the MCP assembly but the field and grid configuration are identical to that used for flight. Using an imaging anode from Surface Science Inc., behind 3 MCPs, images of ions impinging in the target area of the Concentrator can be viewed in real time. Figure 11 shows a Concentrator EM with the MCP imaging assembly in place in the ion beam test facility. Figure 12 shows a detailed drawing of the MCP assembly mounted in the concentrator. Figure 13 shows a typical beam spot as imaged by the MCPs both as the ions enter the test chamber and as they impact the target surface. Spot position comparisons with simulations were made by finding the two-dimensional Gaussian center of each recorded spot and the position of maximum count value in each spot and comparing them with the spot center calculated by SIMION. Figure 14 shows a comparison between MCP image spot maximum and gaussian center fit locations and SIMION calculations.

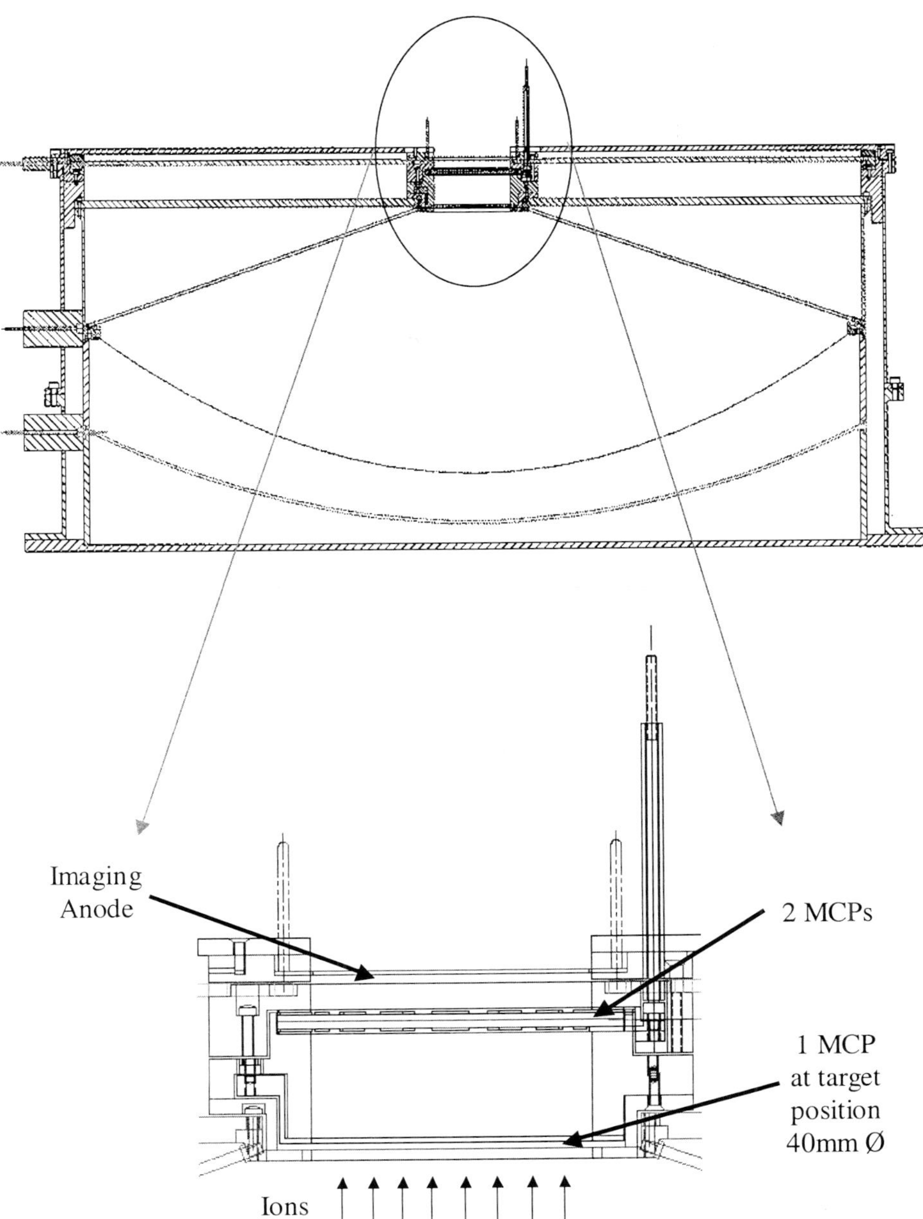

Figure 12. The upper panel shows a cross section of the Concentrator with the MCP imaging assembly installed. The bottom panel shows an enlarged view of the MCP assembly itself. The large vertical structures at the sides of the bottom illustration are the high voltage input pins needed to power the MCP stack.

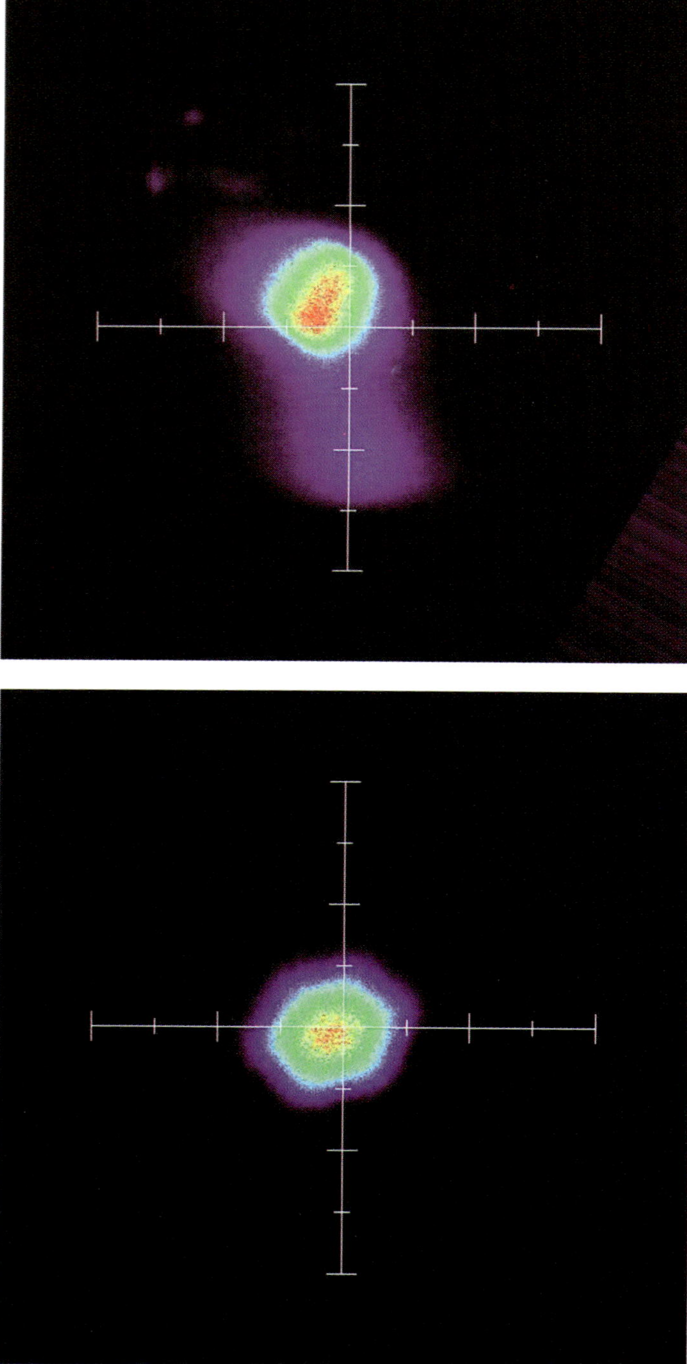

Figure 13. Two ion beam images produced with MCP assemblies. The upper panel shows an image of the beam before entering the Concentrator. The lower panel shows the same beam as it impinges on the MCPs at the target location. The hexagonal shape is due to refraction off the hexagonal etched grid used in the prototype and is only observed when the voltage on the rejection grid is nearly equal to the incident ion energy.

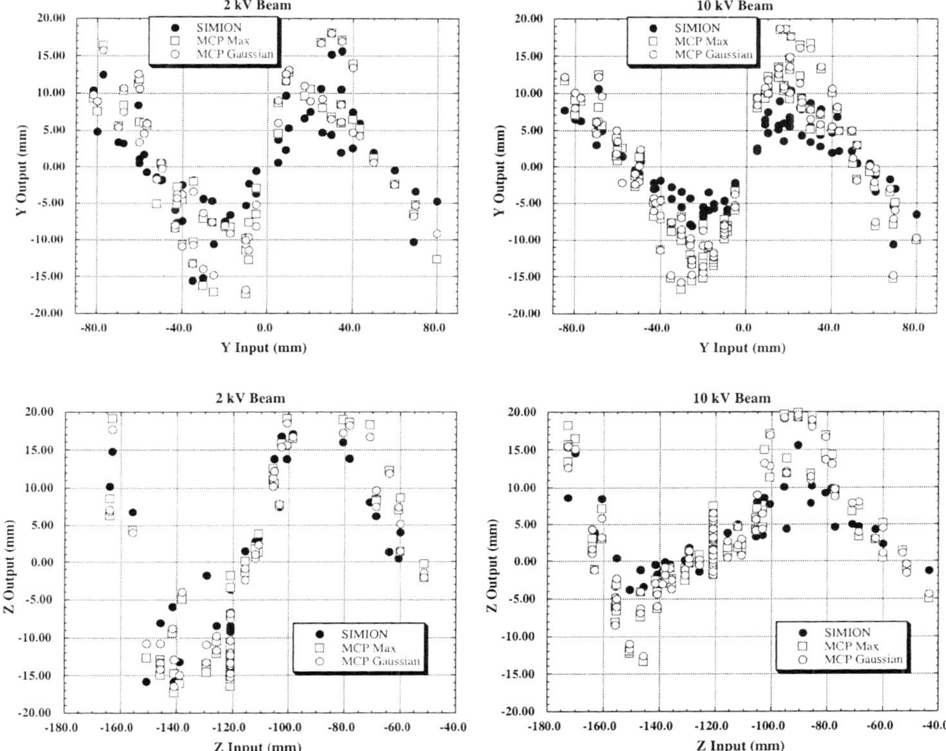

Figure 14. Comparison of the SIMION-predicted location of ion beams on the target and the location determined from the MCP imaging system data vs. the input beam position. The Y and Z-axes are orthogonal directions parallel to the surface of the target. Despite the use of mapping data in the SIMION calculations, there is a systematic under-prediction of the radial spot location.

4.2. Foil Implantation Tests

Mechanical requirements prevented the MCP imaging area from reaching the edge of the target which prevented the MCP imaging method from verifying the Concentrator's operation near the edge of its target and by analogy, the outer portion of its aperture. The outer portion of the Concentrator aperture comprises a major portion of the overall aperture area and also suffers from field variations due to edge effects. The standard methods of eliminating edge effects would require relatively large gaps between the external housing and the grids and other electro-optical elements. This would be an inefficient use of the limited volume and footprint available to the Concentrator onboard the spacecraft. For these areas foil implantation can be used to check the results of simulations in the outer region of the target. The number of data points is small in comparison to that available from MCP testing and this method also requires an understanding of the full implantation and concentration operation including focusing efficiency and backscattering off the

TABLE VII

Comparison of implanted ions with SIMION simulations.

	Portion of target	Measured fraction	SIMION
He[a]	Inner 1 cm	0.64	0.63
	Middle 1 cm	0.31	0.30
	Outer 1 cm	0.06	0.07
Ne[a]	Inner 1 cm	0.99	1.00
	Middle 1 cm	0.00	0.00
	Outer 1 cm	0.01	0.00

[a]Test conducted with EM. He and Ne were implanted with 140 mm straight line scans. For He the scan was perpendicular to and bisecting a radial arc at $R = 121$ mm. For Ne the scan was nearly radial, near a support rib, from $R = 50$ to 190 mm. All scans used normally incident 9 kV beams with mirror and acceleration voltages at ± 10 kV, respectively. All noble gas measurements were corrected for backscattering losses.

target. Foil implantation testing was performed by covering the target area with cleaned aluminum foil which has a very low background content of noble gases. The Concentrator was then exposed to beams of the noble gases He^+, Ne^+, and Ar^+ and the analysis of the implanted noble gases were then compared with the amount expected from implantation time and measurements of the beam made with a Faraday cup and a non-concentrated control sample implanted into another foil. These tests showed good agreement with predictions. Table VII shows a comparison of predictions from the simulations and measurement of implantation in the foils for one implantation test. This demonstrates both a good understanding and agreement with simulation of the Concentrator's optics, mechanical construction, and implantation processes.

4.3. Grid and Surface Mapping

Several times in the Concentrator's development and preparation for flight, its targets cannot be examined directly to verify its continued correct operation. Concentrator operation must also be verified without access to its target in thermal vacuum and solar thermal vacuum testing as well as after final flight cleaning and preparation. The MCPs and foil implantation cannot be used during thermal vacuum tests because the vacuum levels in these chambers are not sufficient. Also, when the Concentrator is finally assembled for flight, the targets in the Concentrator cannot be installed without partial disassembly of the upper portion of the flight instrument. Any misalignment, wrinkling, or damage to grids could be detrimental to the Concentrator's operation and must be detected before flight. After final flight assembly the Concentrator is put in a high vacuum chamber and brought to full voltage to verify that no loose wires have been included and the operation of the

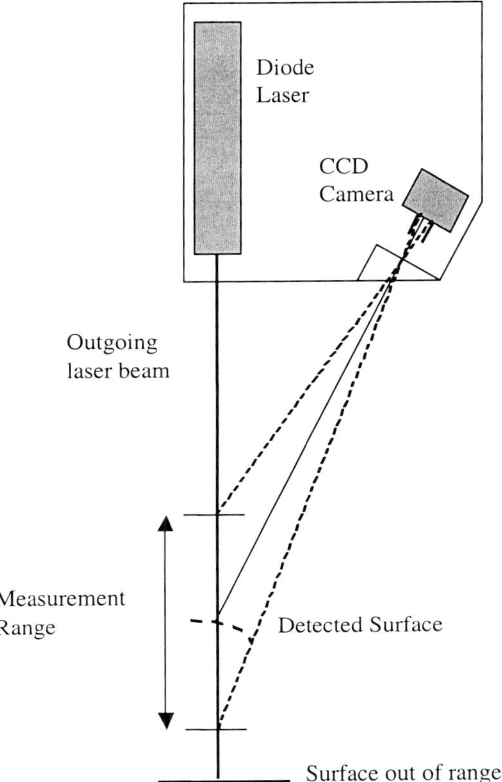

Figure 15. The Concentrator optical mapping facility uses optical micrometers which operate by detecting the diffuse reflection off a surface. Several different micrometers with different measurement ranges were used to select the grids and surfaces of interest on each pass of the mapping facility's translation stages.

power supplies is checked in high vacuum where the self-tests and current monitors for each of the high voltages produced by the flight CEBs can easily be verified. However the Concentrator could not be exposed to any ion beams or other sources that may cause unintentional implantation in the targets. Nevertheless, the FM Concentrator must be verified to be within the performance specifications listed above in a way that does not expose it to any possible contamination or implantation. Since the detailed operation of the Concentrator depends most directly on the exact nature of the surfaces of each of the grids and the mirror electrode, operational factors can be determined by accurately mapping each of the electro-optical surfaces in a Concentrator. Thus a non-invasive method of analyzing the state of the Concentrator's electro-optical surfaces was needed. Laser mapping of the grid, support, and mirror surfaces together with raytracing of the ion optics made it possible to assess the state of the Concentrator without activating it. However, mapping the surfaces of grids which are more than 90% transmissive

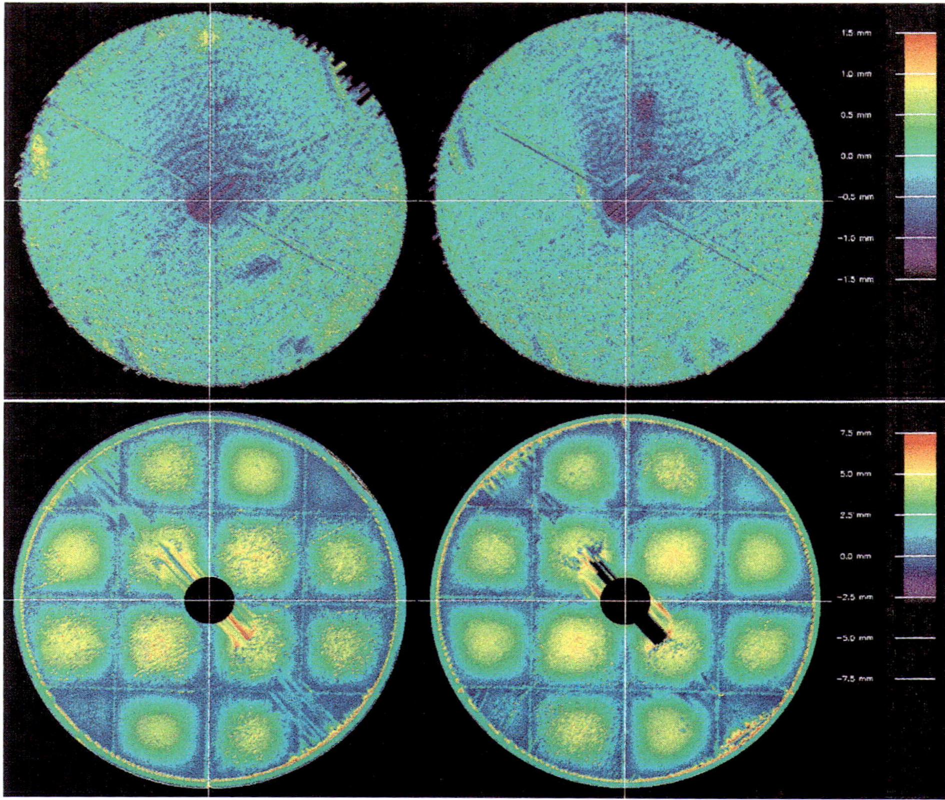

Figure 16. Color coded mapping of deviations from ideal for the FM acceleration grid surface (top) and FM domed grid (bottom) before and after vibration. The hydrogen rejection maps are referenced to a flat plane, while the domed grid maps are referenced to a paraboloid, so the dimples between supports are clearly visible. Only very small changes are visible indicating good dimensional stability for the grids.

in a fully assembled instrument where most surfaces are not directly accessible, required the development of a unique optical mapping facility. Finely focused optical triangulation micrometers built by Precimeter proved capable of locating and determining the distance to a grid wire to within 0.1 mm and to 0.03 mm for solid surfaces. Each electro-optical surface of the Concentrator can be mapped for shape and alignment within tolerances on a 1 mm by 1 mm square grid using these sensors. A mapping facility using a 3-axis optical translation table to automatically raster the optical micrometer over the Concentrator and record the absolute position of each surface was developed. Figure 15 shows a schematic diagram of the laser micrometer's operation in the mapping system. The different focal ranges of a variety of these optical micrometers is coupled with the approximate distance the translation stages are programmed to maintain from the Concentrator so that each surface being mapped is differentiated from the other surfaces for each raster pass. A specification of no more than ±3 mm deviation from the ideal surface figure and no more than 0.3° misalignment with the housing was established using SIMION raytracing. A Concentrator meeting these surface requirements will electro-optically perform within design requirements.

Every effort was made to map Concentrators under all of the various conditions to which they are subject to understand any changes in surfaces that would occur from such environmental factors as heating or vibration. Maps from before and after vibration are shown in Figure 16 and illustrate the excellent dimensional stability of all Concentrator components. Although no micro-meteorite impact tests were performed on a Concentrator, the extremely strong woven grids are difficult to damage relative to other grids of the same open area ratio and do not fail catastrophically when cut because of the energy absorption afforded by the bending of the wires in the weave. Although micrometeoroid impact testing was not performed, the grids are expected to survive such impacts better than other mesh because a single wire can be broken without moving relative to its nearest neighbors or causing the grid to tear.

In addition to performing these critical verifications, the mapping facility was used to measure the added wrinkling expected in the grids when in space. When fully exposed to the Sun, as they will be throughout the collection portion of the mission, the grids wires are expected to be warmer than their support structures. Because of the extremely small diameter of the grid wires, their temperatures cannot be measured with any standard contact thermometer. The extremely rapid cooling of the grid wires when shadowed, requires that the state of the grids be measured while under direct insolation, making infrared temperature measurements very difficult. The mapping facility once again served to provide data on the actual state of the grids. The optical micrometers have a notch filter that allows them to operate in sunlight. The wrinkling of the grids under direct sunlight was measured in the LANL solar thermal vacuum test chamber by a straight-forward modification of the mapping facility which allowed it to map the grids through the chamber window. The measured wrinkling of the grids not only verified acceptable performance of the Concentrator under flight-like conditions but can also be used to estimate the temperature of the grids.

4.4. Instrument Modeling

Actual concentration factors and fractionation depend strongly on the actual solar wind conditions, which differ significantly from what could be achieved in the test chamber. Realistic solar-wind performance of the concentrator could only be modeled rather than duplicated in the lab. Modeling of the concentrator performance under actual solar wind conditions was carried out using a SIMION model as described in Wiens *et al.* (2003). The validity of the SIMION model was determined with data from the prototype, EM, and FM Concentrators by comparing simulations using the laser mapping data with the results of the MCP tests. Fig-

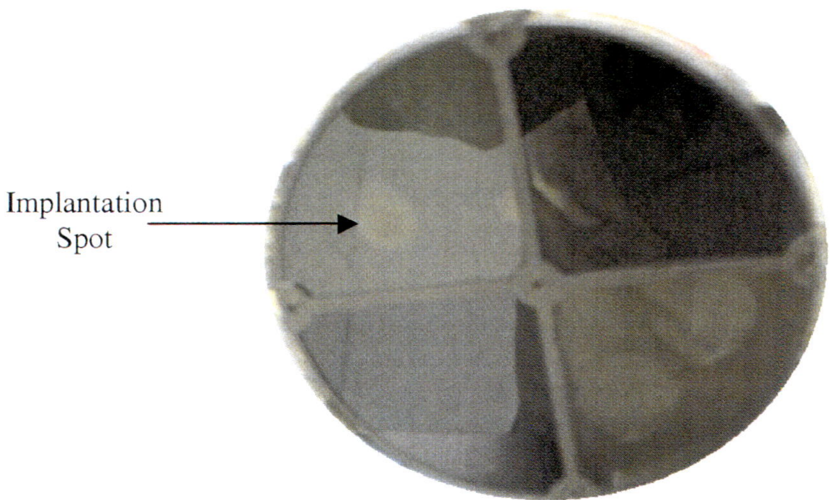

Figure 17. Concentrator target used for implantation test. The location of the implantation is clearly visible in a SiC segment. Several reflections from items on nearby work benches are also visible.

ure 14 shows a comparison between simulated beam positions onto the target given a known beam input location and beam positions as measured by the MCP system. The agreement between the two is good but the simulations consistently underestimate the radius of the beam spot's position on the target. Systematic focusing effects in the MCP optics are believed to be responsible for the discrepancy, and because of good agreement between foil tests and modeling results, the Concentrator focusing is believed to be sufficiently well understood that quantification of the mass fractionation of the ions implanted in the target is below the design requirement. One difficulty with the MCP data is that all 3 flat grids must be changed out to install the MCP system necessitating remapping to determine the exact form of each grid each time a change is made.

4.5. Target Implantation

Target implantation, like foil implantation is a check of the fidelity of the full Concentrator system, design, simulation codes, and operation showing for example that grid transparency is as expected. However target implantation also confirms the complete target methodology from target production and handling to the analysis methods to be used for post-flight scientific investigations. Figure 17 shows a target after an implantation has been performed. This target has been archived for future analysis.

5. Components and Component Level Testing

Several specialized components were developed for the Concentrator because of the unusual demands of the instrument's design. Any component not already proven in flight applications had to undergo rigorous testing beyond the standard tests they endured as part of the Concentrator.

5.1. GRIDS

Extremely strong mesh with the highest possible open area ratio was very important to the overall operation of the Concentrator. Woven mesh was found to be optimal for the Concentrator. This mesh is considerably stronger and more consistent than etched mesh. In order to have sufficient open area, etched mesh must be eroded until it develops weak points in the corners where the etching process invades grain boundaries in the metal. Etched mesh also must be made from a very thin sheet of material so that if the bars of the mesh measure 25 μm along the surface they are generally only $\sim$2.5 μm normal to the surface. Woven mesh wires are the same thickness in both directions and provide some extra resilience because of the slightly wavy pattern induced by weaving. The mesh in the concentrator is made from 400 series stainless steel wire 25 μm in diameter woven on a 20 lines cm^{-1} pitch. Pull tests confirmed additional strength from coldworking of the wires when they are drawn and the added ductility of the woven configuration. Immersion of mounted grids and grid samples in liquid N_2 confirmed that there would be no phase change in the material, which might cause dimensional and strength variations when shadowed during spacecraft maneuvers.

5.2. INSULATOR COATINGS

The alumina high voltage standoffs separating the various Concentrator electrode elements are a possible source of field variations. Sunlight striking electro-optical surfaces within the Concentrator will generate photoelectrons, which in some cases will be drawn toward the insulating standoffs because of the orientation of the electric fields around them. Bare alumina has a bulk resistivity $> 10^{16}\,\Omega$, which is sufficiently high that the standoffs may not be able to bleed off the photoelectrons quickly enough to prevent significant charge accumulation and possible voltage breakdown. This effect is generally small and can be studied under normal laboratory testing. However, in the case of insulators with conducting surfaces nearby which will be in full sunlight, raytracing is necessary to determine if photoelectrons deposited on the insulating surfaces result in increased charging and cause unwanted electro-optical effects. This effect cannot be fully tested under laboratory conditions because it requires simultaneous use of a sun source and an ion beam source, which, for high fidelity, requires the use of different test chambers. The maximum electron flux from the solar irradiation of metallic parts onto the insulators can be determined by first calculating the photoelectron production rates for

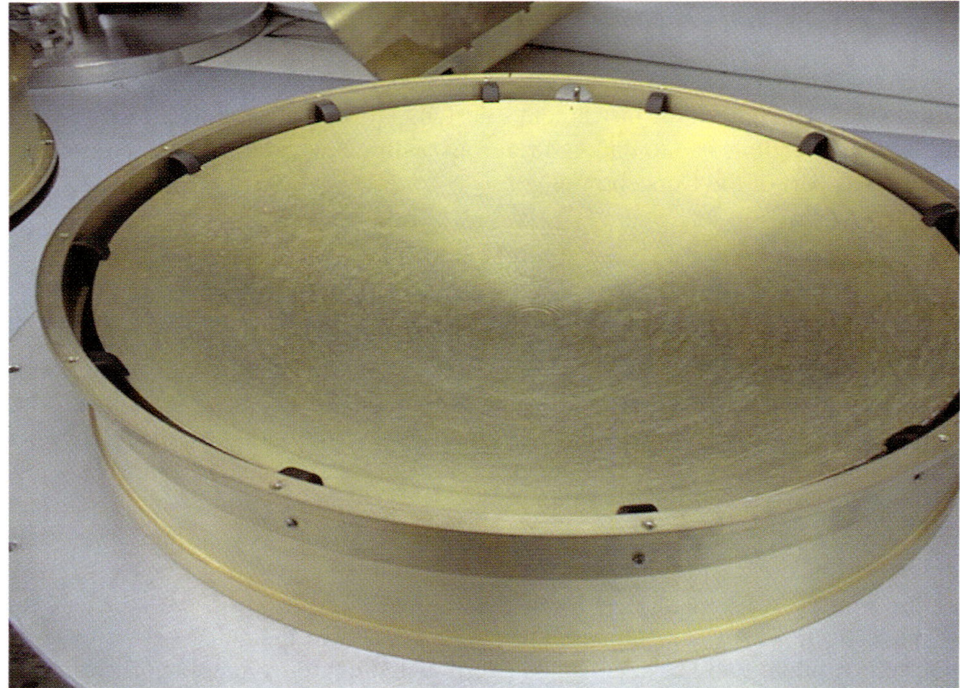

Figure 18. The lower portion of the Concentrator showing the microstepped mirror electrode. The N-doped SiC coated alumina insulators used around the edge of the mirror electrode appear black instead of the normal white of uncoated alumina. The opening in the housing and input post where the high voltage connection is made to the mirror electrode is also visible at the top of the photo.

the conducting surfaces and then performing trajectory simulations to determine transport efficiency of the photo electrons to the standoffs. If particle trajectories are disturbed by charging of the standoffs, resistance of the standoffs must be reduced significantly below their bare-alumina values.

Analysis of this effect showed that the only insulators subject to excessive charging are the mirror electrode's edge standoffs which are in the direct path of the photoelectrons ejected from the outer edge of the domed grid and the central standoff between the target and the target sun shade which is struck by ions which miss the target. Approximately 5 nA of photoelectron current must be discharged by the 12 mirror electrode of standoffs, or ~ 0.5 nA per standoff, which determined that the maximum resistance should be less than 10^{13} Ω per standoff. This is balanced by the need to keep the current drawn from the high-voltage power supply less than the supply limit of 3 μA. This requires that the standoffs have a resistance of at least 10^{10} Ω per standoff. Thus a resistance of about 3×10^{11} Ω per standoff between factors of 1/30 and 30 was desired. For the central standoff the resistance required was between 10^{10} and 10^{14} Ω.

The coating chosen for the standoffs was SiC doped with nitrogen. The nitrogen content in the SiC controls the bulk resistivity of the film and the SiC provides a hard, low-oxygen easily-handled surface. The N content is too low to cause concern over possible target contamination from sputtering of this coating. The film is normally deposited by a discharge sputtering method in an argon environment of about 10 mTorr. By adjusting the partial pressure of nitrogen gas in the discharge environment, the nitrogen doping content can be controlled. For the 2.5 μm-thick film used here, the desired nitrogen partial pressure was empirically determined to be 5% in order to achieve the desired resistivity ($\sim 10^8$ Ω-cm). The coating thickness was determined by the vendor (Technology Assessment and Transfer, Inc.) using profilometry on witness slides. The most critical property of the SiC films is that they do not reduce the resistance between electrodes below the level necessary to operate the high voltage supplies. This was verified during the high voltage functional test of the FM after buildup. The current drawn from the power supplies was monitored to verify that the required level was acceptable. To verify that the standoff resistance is also low enough to bleed off photoelectrons, a representative sample of the coated standoffs was tested to determine the individual resistances of the SiC films. Final resistance of flight standoffs was 5×10^{11} Ω for each mirror electrode standoff and 2×10^{13} Ω for the target standoff.

To determine if these films had the necessary stability for use in flight, lifetime testing of SiC films on glass slides and on alumina disks were placed in an oven at 80 °C while the humidity was held at 100% for a period of 1 month. This simulated conditions prior to launch and is considered a significant over-test of expected environmental conditions. The films were doped with varying percentages of nitrogen partial pressure, ranging from 0 to 40%. Films coated on alumina showed no signs of wear. However to varying degrees, all of the samples on glass slides showed peeling, except for a pure SiC film. Because the films on alumina represent the standoff configuration and the glass slides do not, the failure of the films on glass slides was not considered to be of concern. Two samples of SiC coated on alumina, one at 7% nitrogen partial pressure and the other at 16%, were next exposed to UV radiation from a xenon lamp to simulate solar irradiation. The exposure dosage (intensity × time) was 10 times the UV exposure expected over the mission lifetime. The 7% sample showed no signs of wear, whereas the 16% sample appeared to 'fade', as if the coating had been being ablated. The fading was particularly significant around the edges of the piece. The adherence of the films does not appear to be affected. Both coatings still pass the 'tape' test and do not abrade while being rubbed with a latex-gloved finger. The cause of the fading is unknown. The exposure was carried out in air, and more importantly, in a fairly significant ozone environment (the ozone being created by the UV lamp). Thus the fading could either be due to a direct reaction with the UV radiation or a reaction involving the presence of air/ozone. The UV exposure test was considered to be successful for the Concentrator standoffs because the nitrogen partial pressure for coating the flight standoffs was 5%, which is below the 7% partial pressure coating conditions used for the sample that showed no signs of wear. The integrity of the coatings is expected to increase at lower nitrogen pressures as their composition becomes more like pure SiC. Thus, flight standoff coatings have wear properties that are as good as or better than the 7% sample. Neither the mirror standoffs nor the central target support standoff are directly illuminated in normal Concentrator operation. Finally, the tested exposure was a factor of 10 in excess of the expected dose, and in the unlikely event that some ablation occurs, it should be very much less than that in our testing. Figure 18 shows the coated mirror electrode standoffs.

6. Operation

The Concentrator voltages are adjusted in real time onboard to react automatically to changes in the solar wind. However, Concentrator voltages should not be changed too rapidly and consequently have limited deltas on existing potentials. Defining $\Delta V_? = (V_?)_n - (V_?)_{n-1}$ where $(V_?)_n$ is the desired new voltage and $(V_?)_{n-1}$ is the readback voltage of the present CEB setting then $|\Delta V_?| \leq 500$ V. If the desired change is more than 500 V, the CEB will catch up after 2 or more cycles of 5–30 (nominally 30) s. The acceleration voltage will remain at −6.5 kV during normal operation. The mirror and hydrogen-rejection grid voltages are set based on the peak in the proton energy distribution, Ep. Ep is a running average of the last three consecutive measurements of Ep defined as

$$Ep = (Ep_n + Ep_{n-1} + Ep_{n-2})/3$$

to help smooth out any noise from individual measurements. The mirror voltage is

$$V_M = Ep^*4.32$$

as described in the Overview section. The hydrogen rejection-grid is set at

$$V_H = Ep^*R,$$

where R is given in Table IV. The operational settings of the Concentrator are archived along with the solar wind moments from the monitors, so that the Concentrator target samples will be understandable in the context of the solar wind conditions experienced during the mission.

7. Summary

The extremely demanding scientific and operational requirements placed on the Concentrator have all been met. Table II gives a complete list of performance requirements for the flight Concentrator. Many types of testing and characterization were used to verify that all of these criteria have been met. In order of the listing in Table II here are the results of these tests and the expected performance of the Concentrator.

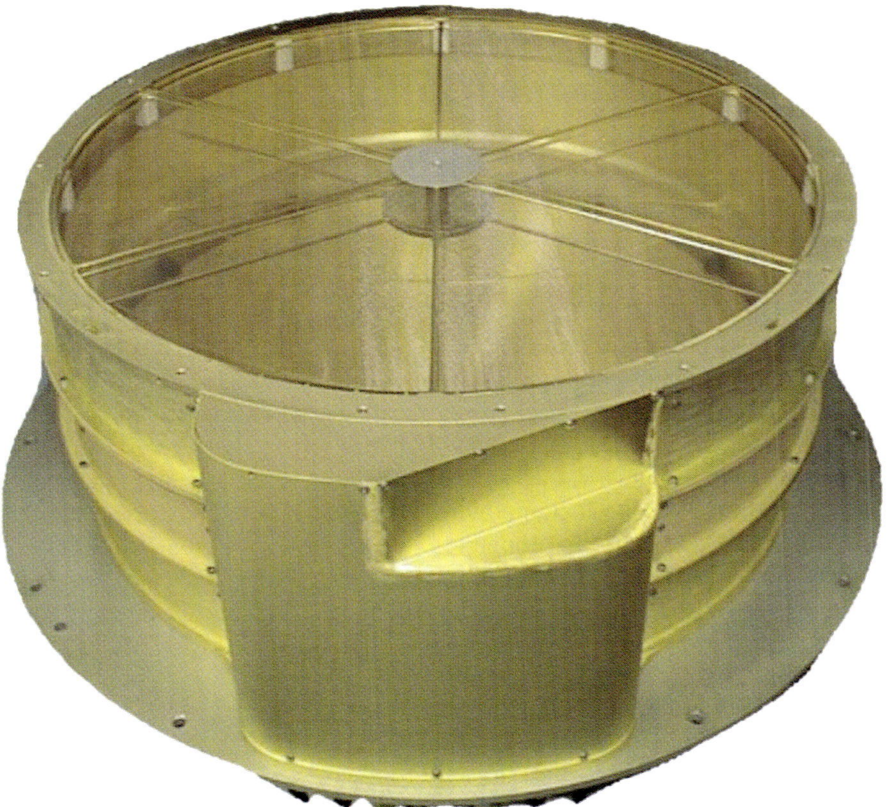

Figure 19. Final view of the FM Concentrator after gold coating, high voltage testing, and mapping in preparation for final mating with the science canister and spacecraft. The welded cover of the high voltage input section is closest to the camera.

1. The concentration factor varies for the type of solar wind encountered and reaches a maximum of 22.8 for low-speed solar wind. The worst case is 16.8 for high speed solar wind which occurs $\sim 1\%$ of the time. The concentration factor is calculated as follows:

   Geometric Factor $=$ (Aperture Radius)^2/(Target Radius)^2 $=$ $(20\text{ cm})^2/(3.1\text{ cm})^2 = 41.6$

   GridTransmission $=$ (GridOpenAreaRatio)$^{(\text{Number of grids traversed})}$ $= (0.9044)^5 = 0.605$

   Fraction Hitting Target $= 0.669$ to 0.907 from simulations

   Concentration Factor $=$ Geometric Factor $*$ Grid Transmission $*$ Fraction Hitting Target $= 16.8$ to 22.8.

The fraction of ions hitting the target will be determined by SIMION for the solar wind conditions encountered. Thus the expected integrated concentration factor is over 20 but will depend on the exact solar wind conditions encountered.

2. Sufficient target area is needed to allow several analyses of the target material and archival of $\geq 50\%$ for future analysis. This required a target area of ≥ 15 cm^2. The measured target active area in the Concentrator is 25.9 cm^2. While this is relatively large, the concentration of ions in the target decreases as a function of radius so that the most desirable portions of the target will be the inner ~ 15 cm^2. Figure 6 shows the target assembly including the alternating wedges of SiC and diamond. These different materials have different background contamination levels of the elements and isotopes of interest and the separate analyses and background levels should allow for more verifiable results.
3. The Concentrator must focus ions with widely varying charge state and velocity distributions into its target without introducing unknown variations and, thus, analysis errors, in the implanted $^{17}O/^{16}O$ ratio different from that in the solar wind by greater than 0.1%. Implantation variations can be calculated with SIMION before flight to this level and will be checked after flight with Ne^+ implantation levels. Details of the simulations used to verify this level of operation are given in Wiens *et al.*, 2003.
4. The Concentrator target temperature must not exceed 250 °C to avoid damage to the target material and release of the implanted species. The calculated worst-case temperature is 230 °C.
5. To further avoid damage to the target from hydrogen buildup, 90% of Solar wind proton fluence must be prevented from reaching the target. Figure 4 shows that with the planned operational methodology the current design rejects $\sim 93\%$.
6. Concentrator cleaning, coating, handling, and final cleaning methods have been verified to have a surface contamination by C, N, O below 10^{15} atoms cm^{-2} under ultra high vacuum ($< 10^{-8}$ torr) conditions at 200 °C. The level of contamination was continually monitored in all operations before flight by the use of witness plates. In addition, the Concentrator underwent a final cleaning and gold coating after all testing was complete and just before flight to eliminate any possibility of contamination during handling. Both of these operations had been previously analyzed and found to meet the contamination design specification. Once in flight, contaminates do not implant and the outer layer of the targets will be removed before analysis of the implanted solar wind is performed.

The FM Concentrator in its final configuration for flight is shown in Figure 19. We look forward to the return of the Concentrator and its targets. Upon retrieval the Concentrator will be returned to Los Alamos National Laboratory for analysis of its operation with the mapping facility, the targets will be removed under ultra-clean conditions, and then the Concentrator will be tested in the ion beam facility. If the midair capture of the sample-return canister fails we expect that there will be some damage to the Concentrator that may make mapping and operation in the ion beam impossible but we expect to retrieve the target pieces intact.

8. Concentrator Design Parameters

Overall diameter	46 cm
Overall height, excluding CEBs	21 cm
Working aperture	40 cm
Target diameter	6.2 cm
Focal length, mirror electrode	21.7 cm
Focal length, domed grid frame	14.0 cm
Mirror grid stand-off from electrode	$\geq$ 2.5 cm
H-grid to ground grid stand-off	0.6 cm
Acceleration grid to H-grid stand-off	1.8 cm
Domed Grid to Acceleration Grid at Center	14.35 cm
Domed Grid to Target at Center	13.85 cm
Electrode microstep height	100 microns
Concentrator mass, excluding CEBs	6.0 kg
HVPS mass, (CEBs, each of two)	1.4 kg
Maximum allowable target temperature	250 °C
CEB operating thermal range	−10/+77 °C
CEB survival thermal range	−15/+77 °C
Ground Grid Aperture	41.6 cm
Domed Grid Aperture	40.0 cm
Mirror Electrode Diameter	41.9 cm
Angle of electrode at edge, relative to normal	25.8°
Offset angle relative to canister	0°
Acceleration potential	−6.5 kV
H grid potential range	0.1 to 3.5 kV
Mirror Electrode potential range	2.0 to 10.0 kV
Nominal H grid potential	$\sim$ 1.3*Ep
Nominal mirror potential	4.32*Ep
H grid steps	0.855 V linear
Mirror electrode steps	2.44 V linear
Desired solar wind velocity range	300–800 km s^{-1}
Desired solar wind energy range	0.47–3.34 keV amu^{-1}
Desired solar wind m/q range	2.0–3.6
Mirror penetration for (hypothetical) ions normally incident at center (% of distance from grid to electrode)	
$m/q = 2.0(^{16}O{+}8)$:	83.8%
$m/q = 3.6(^{18}O{+}5)$:	95.0%

Acknowledgements

The authors would like to acknowledge NASA contract number W-19,272 for supporting this work. The design, development, and construction of the Concentrator would not have been possible without the help of many individuals. The authors would like to thank the following outstanding individuals: Diane Albert, Frank Ameduri, Matt Anderson, Richard Bramlett, Randy Edwards, Brian Henneke, James Lake, and Stacy Rupiper of Los Alamos National Laboratory; Dennis Guerrero, Irene Arevalos, Greg Dirks, Jeff Roese, Toby Stecklein, Jack Taguiam, Syrrel Rogillio, and James Sanders of the Southwest Research Institute; Chet Sasaki, Virgil Mireles and Don Sevilla of the Jet Propulsion Laboratory; Lada Adamac of Caltech; Allen Dorn of Screen Technology Group; Jerry Spieckerman of Marketech, Inc.; Steve Good of Reynolds Industries, Inc.; Kenneth Bedard formerly of Precimeter, Inc.; and David King and Joseph Maciejewsky of Technology Assessment and Transfer, Inc., Wear Sciences Division.

References

Barraclough, B. L., Dors, E. E., Abeyta, R. A., Alexander, J. F., Ameduri, F. P., Baldonado, J. R., Bame S. J., Casey P. J., Dirks, G., Everett D. T., Gosling, J. T., Grace, K. M., Guerrero, D. R., Kolar, J. D., Kroesche, J., Lockhart, W., McComas, D. J., Mietz, D. E., Roese, J., Sanders, J., Steinberg, J. T., Tokar, R. L., Urdiales, C., and Wiens, R. C.: 2003, 'The Plasma Ion and Electron Instruments for the Genesis Mission', *Space Sci. Rev.*, this volume.

Bochsler, P.: 2000, 'Abundances and Charge States of Particles in the Solar Wind', *Rev. Geophys.* **38**(2), 247–266.

Bochsler, P. and Geiss, J.: 1989, 'Composition of the Solar Wind', in J. H. Waite, Jr., J. L. Burch, and R. L. Moore (eds.), *Solar System Plasma Physics, Geophysical Monograph 54* pp. 133–141.

Bodmer, R. and Boschler, P.: 2000, 'Influence of Coulomb Collisions on Isotopic and Elemental Fractionation in the Solar Wind Acceleration Process', *J. Geophys. Res.* **105**(A1), 47–60.

Bühler, F., Eberhardt, P., Geiss, J., Miester, J., and Signer, P.: 1969, 'Apollo 11 Solar Wind Composition Experiment: First Results', *Science* **166**, 1502–1503.

Bühler, F., Geiss, J., Miester, J., Eberhardt, P., Huneke, J. C., and Signer, P.: 1966, 'Trapping of the Solar Wind in Solids', *Earth Planetary Sci. Lett.* **1**, 249–255.

Burnett, D. S., Barraclough, B. L., Bennett, R., Neugebauer, M., Oldham, L. P., Sasaki, C. N., Sevilla, D., Smith, N., Stansbery, E., Sweetnam, D., and Wiens, R. C.: 2003, 'The Genesis Discovery Mission: Return of solar matter to Earth', *Space Sci. Rev.*, this volume.

Clayton, R. N.: 1993, 'Oxygen Isotopes in Meteorites', *Ann. Rev. Earth Planetary Sci.* **21**, 115–149.

Collier, M. R., Hamilton, D. C., Gloeckler, G., Ho, G., Bochsler, P., Bodmer, R., and Sheldon, R.: 1998, 'Oxygen 16 to Oxygen 18 Abundance Ratio in the Solar Wind Observed by WIND/MASS', *J. Geophys. Res.* **103**, 7.

Geiss, J., Eberhardt, P., Signer, P., Bühler, R., and Meister, J.: 1969, 'The Solar Wind Composition Experiment', Section 8 of *Apollo 11 Preliminary Science Report*, NASA SP-214, pp. 183–186.

Geiss, J., Eberhardt, P., Bühler, R., Meister, J., and Signer, P.: 1970, 'Apollo 11 and 12 Solar Wind Composition Experiments: Fluxes of He and Ne Isotopes, *J. Geophys. Res.* **75**, 5972–5979.

Geiss, J., Bühler, R., Cerutti, H., Eberhardt, P., and Filleux, Ch.: 1972, 'Solar Wind Composition Experiment', *Apollo 14 Preliminary Science Report*, NASA SP-315, 14-1–14-10.

Harris, M. J., Lambert, D. L., and Goldman, A.: 1987, 'Carbon and Oxygen Isotope Ratios in the Solar Photosphere', *Monthly Notices Roy. Astron. Soc.* **224**, 237.

Jurewicz, A. J. G., Burnett, D. S., Wiens, R. C., Friedmann, T. A., Hays, C. C., Hohlfelder, R. J., Nishiizumi, K., Stone, J. A., Woolum, D. S., Becker, R., Butterworth, A. L., Campbell, A. J., Ebihara, M., Franchi, I. A., Heber, V., Hohenberg, C. M., Humayun, M., McKeegan, K. D., McNamara, K., Meshik, A., Pepin, R. O., Schlutter, D., and Wieler, R.: 2003, 'Overview of the Genesis Solar-Wind Collector Materials', *Space Sci. Rev.*, this volume.

Kallenbach, R., Geiss, .J., Ipavich, F. M., Gloeckler, G., Bochsler, P., Gliem, F., Hefti, S., Hilchenbach, M., and Hovestadt, D.: 1998, 'Isotopic Composition of Solar Wind Nitrogen: First in situ determination with the CELIAS/MTOF spectrometer on board SOHO', *Astrophys. J.* **507**(2), L185–L188.

Kallenbach, R., Ipavich, F. M, Bochsler, P., Hefti, S., Hovestadt, D., Grunwaldt, H., Hilchenbach, M., Axford, W. I., Balsiger, H., and Burgi, A.: 1997, 'Isotopic Composition of Solar Wind Neon Measured by CELIAS/MTOF on board SOHO', *J. Geophys. Res. Space Phys.* **102**, 26895–26904.

McComas, D. J. *et al.*: 1998, 'Solar Wind Concentrator', in R. F. Pfaff, J. E. Borvsky, and D. T. Young (eds.), Measurement Techniques for Space Plasmas, AGU Monograph 102, pp. 195–200.

Neugebauer, M., Steinberg J. T., Tokar R. L., Barraclough B. L., Dors E. E., Wiens R. C., Gingerich D. E., Luckey D., and Whiteaker D. B.: 2003, 'Genesis On-Board Determination of the Solar Wind Flow Regime', *Space Sci. Rev.*, this volume.

Pepin, R. O.: 1991, 'On The Origin And Early Evolution Of Terrestrial Planet Atmospheres And Meteoritic Volatiles', *Icarus* **92**(1), 2–79.

Signer, P., Eberhardt, P., and Geiss, J.: 1965, 'Possible Determination of the Solar Wind Composition', *J. Geophys. Res* **70**(9), 2243–2244.

Wiens, R. C., Huss, G. R., and Burnett, D. S.: 1999, 'The Solar Oxygen-Isotopic Composition: Predictions and Implications for Solar Nebula Processes', *Meteorol. Planetary Sci.* **34**, 99.

Wiens, R. C., Neugebauer, M., Reisenfeld, D. B., Moses, R. W., Jr., and Nordholt, J. E., Burnett, D. S.: 2003, 'Genesis Solar Wind Concentrator: Computer Simulations of Performance under Solar Wind Conditions', *Space Sci. Rev.*, this volume.

Wimmer-Schweingruber, R. F., Bochsler, P., and Gloeckler G.: 2001, 'The Isotopic Composition of Oxygen in the Fast Solar Wind: ACE/SWIMS', *Geophys. Res. Lett.* **28**(14), 2763–2766.

GENESIS SOLAR WIND CONCENTRATOR: COMPUTER SIMULATIONS OF PERFORMANCE UNDER SOLAR WIND CONDITIONS

ROGER C. WIENS[1], MARCIA NEUGEBAUER[2], DANIEL B. REISENFELD[1], RONALD W. MOSES, JR.[3], JANE E. NORDHOLT[4] and DONALD S. BURNETT[5]

[1]*Space & Atmospheric Sciences, NIS-1, MS D466, Los Alamos National Laboratory, Los Alamos, NM 87545, U.S.A.*

[2]*Jet Propulsion Laboratory, MS 169-506, 4800 Oak Grove Drive, Pasadena, CA 91109, U.S.A.**

[3]*Theoretical Division, T-3, MS B216, Los Alamos National Laboratory*

[4]*P-21, MS D454, Los Alamos National Laboratory, U.S.A.*

[5]*Geological and Planetary Sciences, MS 100-23, California Institute of Technology, 1201 E. California Boulevard, Pasadena, CA 91125, U.S.A.*

(Author for correspondence, e-mail: rwiens@lanl.gov)

Received 15 December 2001; Accepted in final form 22 May 2002

Abstract. The design and operation of the Genesis Solar-Wind Concentrator relies heavily on computer simulations. The computer model is described here, as well as the solar wind conditions used as simulation inputs, including oxygen charge state, velocity, thermal, and angular distributions. The simulation included effects such as ion backscattering losses, which also affect the mass fractionation of the instrument. Calculations were performed for oxygen, the principal element of interest, as well as for H and He. Ion fluences and oxygen mass fractionation are determined as a function of radius on the target. The results were used to verify that the instrument was indeed meeting its requirements, and will help prepare for distribution of the target samples upon return of the instrument to earth. The actual instrumental fractionation will be determined at that time by comparing solar-wind neon isotope ratios measured in passive collectors with neon in the Concentrator target, and by using a model similar to the one described here to extrapolate the instrumental fractionation to oxygen isotopes.

1. Introduction

The Solar-Wind Concentrator is designed to provide a high-fluence sample of the solar wind, particularly of elements in the mass range between 4 and $\sim$ 28 amu, by focusing a large cross section of solar-wind ions onto a small target. The primary purpose of the Concentrator is to obtain a sufficient sample of solar-wind oxygen to enable a high-precision isotopic measurement in spite of the ubiquitous oxygen backgrounds in terrestrial materials. Such a measurement may reveal the cause of the oxygen isotopic heterogeneity among solar-system objects (Wiens *et al.*, 1999). A key assumption in this endeavor is that the solar-wind isotopic composition is

*Currently at Lunar and Planetary Laboratory, University of Arizona Space Sciences Building #92, 1629E. University Avenue, Tucson AZ 85721-0092 USA

Space Science Reviews **105:** 601–625, 2003.

representative of the isotopic composition of the solar photosphere for elements heavier than helium. The Genesis mission will measure the isotopic composition of other elements, such as Ne, Mg, or Si, to unprecedented precision in three different solar-wind regimes–coronal hole, interstream, and coronal mass ejections. If the isotopic compositions of these regimes are identical, the solar wind isotopic composition will be assumed to be representative of the solar photosphere. Isotopic differences would complicate interpretations of the solar composition, though information on isotopic fractionation of heavy elements between regimes would be valuable input to solar-wind acceleration models.

The Genesis Solar-Wind Concentrator is a unique instrument in the magnitude of the concentration achieved (averaging a factor of 20 $\times$), the fidelity required in understanding its mass fractionation characteristics (instrumental fractionation needs to be known to one permil, that is, $\pm 0.1\%, 2\sigma$), and in the fact that the instrument focuses ions onto a solid target to be analyzed on the ground after the spacecraft returns, rather than having an active detector. Most solar-wind instruments to date have relatively small apertures, usually well under 1 cm^2, with the largest around 100 cm^2. These instruments can be tested using narrow ion beams that are easily produced in the laboratory. By contrast, the Genesis Concentrator has an aperture of nearly 1300 cm^2. It is practically impossible to reasonably simulate the solar wind with an ion beam covering such a large area. The CASYMS system at the University of Bern comes the closest, as it has a $\sim$0.6 m. diameter plasma beam (Ghielmetti *et al.*, 1983). However, the uniformity of this beam over the 40 cm diameter aperture of the Concentrator is not sufficient to accurately characterize the Concentrator optics to the desired degree. Secondly, the charge states obtainable in plasma test chambers only approach those in the solar wind if an electron cyclotron resonance (ECR) ion source is used, and no such sources are available in tandem with a wide beam. Even with high charge state capabilities, the distribution of charge states, angles, and velocities cannot be duplicated together. Finally, even if a large, high fidelity solar wind-like beam were to be achieved, verification of the Concentrator optics would be very time-consuming because each target analysis is a lengthy process using mass spectrometry, e.g., secondary ion mass spectrometry, and requiring significant isotopic calibration.

For these reasons it was decided to test the solar-wind-condition performance of the Concentrator by using an ion simulation code after verifying that the code gave results predicted by laboratory ion beam conditions. Laboratory validation of the simulations is described in the companion paper on the Concentrator design (Nordholt *et al.*, 2003). Here we describe the Concentrator model and solar-wind conditions used in the simulation code, and give the results of the simulations. Concentrator components mentioned here are defined and illustrated in Nordholt *et al.* (2003), which may be needed for cross-reference. Following a description of the computer model in the next section of this paper, the various solar wind parameters used in the tests are described in Section 3. Specific interactions between the instrument and the solar wind are described in Section 4. Finally, the results, in terms of ion patterns on the target, are given and discussed in Section 5.

While the simulations described here predict that the Concentrator will give the desired performance, these simulations are clearly not the last word on Concentrator performance. Solar-wind oxygen compositions determined from Concentrator samples will rely on corrections for instrumental (e.g., Concentrator) fractionation based on comparison of solar-wind neon isotopic compositions measured in passive collectors with neon isotopic compositions in the Concentrator target. The instrumental fractionation specific to oxygen m/q states will then be determined by re-running the simulations described here for both oxygen and neon under the actual conditions encountered and archived during the Genesis collection period. This will require a good understanding of charge state conditions during this time period, obtainable from other spacecraft instruments such as ACE SWICS.

The simulations described below focus on determining mass fractionation based on mass per charge (m/q) rather than the more familiar energy per charge (E/q) because of the way the Concentrator is operated in space. Because it is impossible for an electrostatic concentrator with a fixed voltage to operate with high efficiency over the large range in solar-wind energies, the Concentrator voltages are adjusted to continuously track the solar wind proton energy based on real-time data from the Genesis ion monitor (Barraclough *et al.*, 2003). By continuously adjusting the voltages, the energy dependence is removed to zeroth order, leaving a dependence on m/q. There are still performance variations as a function of velocity, so mass fractionation estimates are obtained by using reasonable charge state distributions for the solar wind at different velocities, as described in detail below.

2. Concentrator Simulation Model

SIMION 7.0 (D. A. Dahl, Bechtel Idaho) was chosen for Concentrator modeling. While there exist a number of ion-trajectory programs, they tend to be written with electrostatic or magnetic sector spectrometers in mind. Their geometries are quite specific and very different from the Concentrator, with its large field regions. SIMION is sufficiently general that the Concentrator could be successfully modeled. SIMION employs 3-dimensional arrays of points in space over which it solves the Laplace equation. It then 'flies' ions through these arrays, recording their impact positions on solid surfaces.

The simulation model incorporates nearly all of the components used in the actual instrument. The model is shown in Figure 1. Briefly, the major ion optical elements are, in the order encountered by the ions, a grounded grid across the aperture, followed by a proton rejection grid, which is positively biased to reject $> 90\%$ of the protons, minimizing ion radiation damage at the target. The voltage of the grid is adjusted continuously in flight to levels relative to the peak of the real-time proton energy distributions (E_p), as will be described in more detail below.

The third optical element is the acceleration grid, which is biased at −6.5 kV to accelerate the remaining ions, thereby straightening their trajectories and providing more energy for deeper implantation in the target. After the acceleration grid, the ions pass through a field-free region bounded on the other end by a parabolic domed grid. The mirror electrode is positively biased, again based on real-time solar wind energy data as described below. Although parabolic in overall shape, the curvature actually consists of 0.1 mm steps so as to act on photons as a reverse Fresnel lens to avoid focusing sunlight on the target, heating it excessively. The ions are reflected by parabolically shaped equipotentials between the domed grid and mirror electrode. The ions pass back through the domed grid and are focused on the target.

The simulation model was limited by software and memory to less than fifty million array points. Achieving the maximum resolution within this limitation required economizing array points by using symmetry and by eliminating portions of the model that were not necessary. Near the flat grids and the mirror electrode (Figure 1), the geometry of the outer housing has a significant influence on the trajectories of ions near the edge, so the ion optic model must extend all the way out to ∼23 cm radius. If the housing cutout for the high voltage bus bar is ignored (a 2.5 cm diameter hole in the housing that is level with the lip of the mirror electrode; cf., Figure 18 of Nordholt *et al.*, 2003), advantage can be taken of two axes of symmetry, so that the model need only cover one quarter of the Concentrator. Additionally, the region between the accelerator grid and the domed grid is field free, and therefore needs not be modeled with array points. To take advantage of this field-free region in the center of the instrument, the model is broken into two 'instances', or portions containing array points. As shown in Figure 1, instance 1 contains the flat grids and target; instance 2 contains the domed grid and mirror electrode. Ion trajectories start inside of instance 1 just above the ground grid. Once the ions leave instance 1 they coast with no forces acting on them until they enter instance 2 or leave the simulated region. Because the portion in between instances is not modeled, the Accelerator Can, which is the inner wall of the Concentrator in the field-free region, is missing from the simulation model. Most of the ions that would normally strike the Accelerator Can walls are lost from the simulation by exiting between the instances. A few of the ions that would normally strike the Accelerator Can end up outside of the Accelerator Can radius but inside the radius of the grounded housing. To keep these from re-entering the simulation model between the Accelerator Can and the housing, which is physically impossible in the actual Concentrator, a series of narrow rings, providing the desired voltage gradient, were added to the simulation model in this region (Figure 1). The channels between rings of different voltages are too narrow and long to allow ions to pass. The rings have no effect on ions flying in the main portion of the Concentrator.

For this design, the limitations on total array points allow for a maximum array point spacing of 1.5 per mm. The upper instance has 5.3 million points, and the lower instance has 17.7 million points. It would be possible to decrease the array point spacing by a few percent with some loss in computing speed, however an

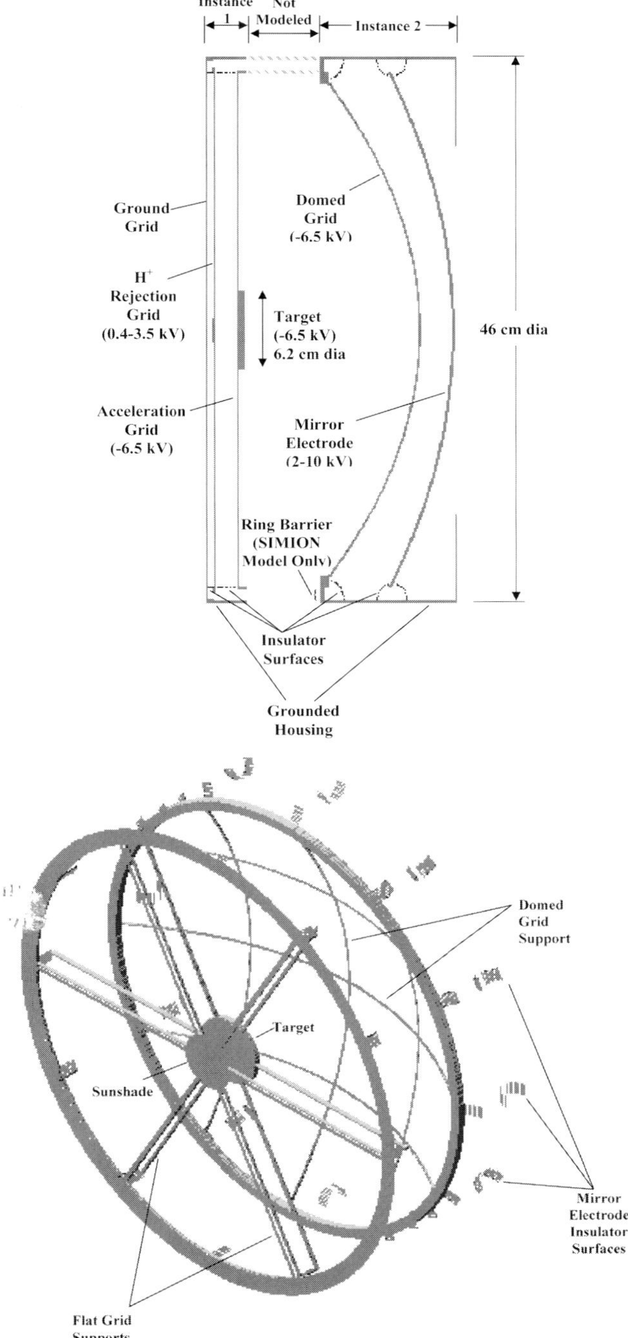

Figure 1. SIMION software model of the solar-wind concentrator. (a) Cross section through the axis of the instrument. (b) Isometric view with housing, grids, and mirror electrode removed to show grid supports, target, sunshade, and insulators. The 3-D model uses an array of 23 million points. Array point spacing is 0.67 mm. The field-free region between Instance 1 (flat grids and target) and Instance 2 (domed grid and mirror electrode) does not need to be modeled. Compare with Figures 2 and 3 of Nordholt *et al.* (2003).

array point spacing of 2.0 per mm would result in over fifty million points, the maximum allowed by the program. The 0.67 mm resolution in the set-up described above is somewhat coarse, but still sufficient for modeling the Concentrator performance as long as certain exceptions are taken into account. In particular, the high transparency grids can only be modeled as smooth membranes rather than individual grid wires. At 0.025 mm, the diameter of the grid wires is far smaller than the model resolution. The generally slight scattering of ions due to passage close to grid wires is therefore treated separately, as described in Section 4.2. Secondly, the mirror electrode microstepping is at 0.1 mm intervals, only a few times smaller than the computer model steps. This difference does not present a problem within the normal range of ion energies input to the Concentrator, for which ions remain much further from the surface.

SIMION does not directly model voltage gradients across insulators. However, in the Concentrator design the insulators are too near the active regions to be ignored. The insulators were therefore modeled as a series of equipotential bars perpendicular to the voltage gradient, as shown in Figure 1. This is certainly reasonable for the slightly-conductive coated insulators supporting the mirror electrode, and should be reasonable for the other insulators, which are uncoated alumina (Nordholt *et al.*, 2003).

3. Solar Wind Parameters

The parameters of the solar wind ions input into the model are critical to obtaining realistic results. This section describes these parameters. The ones considered here are charge state, bulk velocity, angle, and thermal Mach number. For some parameters, namely charge state and angular distribution, the same values were used for nearly all runs, based on grand averages of several years of solar wind data. For bulk flow velocity a number of runs were executed to cover the range of solar wind speeds. For all runs a Monte Carlo spatial distribution was used covering a circular area extending to a radius of 20.8 cm, slightly larger than the 20.6 cm radius of the ground grid, with ions started 1 mm above the grid.

3.1. Charge State Distribution

For oxygen, only charge states +5 to +8 were considered, as lower charge states of actual solar wind constitute well under 1% at all velocities (Wimmer-Schweingruber *et al.*, 1998; Geiss *et al.*, 1992). The charge state distribution used for modeling is given in Table I. Initial modeling used the low-speed solar wind distributions given in Geiss *et al.* (1992). Final calculations used oxygen 7/6 charge state ratios from ACE SWICS data taken between February 6, 1998 and July 1, 2000 for solar wind < 700 km s^{-1}. For 700–800 km s^{-1}, where ACE statistics are poor, Ulysses SWICS data were used. Distributions of the minor ions +5 and +8 were then

TABLE I
Average oxygen charge state distribution used in the Concentrator simulations.

Charge State	Initial modeling	< 400 km s^{-1}	400–500 km s^{-1}	500–600 km s^{-1}	600–700 km s^{-1}	700–800 km s^{-1}
+5	0.01	0.004	0.004	0.0040	0.0045	0.0050
+6	0.67	0.641	0.727	0.8017	0.8360	0.9739
+7	0.24	0.292	0.229	0.1785	0.1550	0.0210
+8	0.08	0.063	0.040	0.0158	0.0045	0.0001
Total	1.00	1.00	1.00	1.00	1.00	1.00
Data source	Geiss *et al.*, 1992	ACE $N = 6571$	ACE $N = 6997$	ACE $N = 3000$	ACE $N = 974$	Ulysses $N = 26006$

calculated for the given 7/6 ratio based on equilibrium thermal models (Summers, 1972; Nahar, 1999).

The ion energy per charge (E/q) ratio strongly affects the degree to which ion trajectories are straightened by the acceleration grid and the depth of penetration of the ion into the region between the domed grid and the mirror electrode for a given Concentrator voltage setting. For a given isotope, adjusting the mirror voltage as a function of E_p largely compensates the energy factor at the mirror (Nordholt *et al.*, 2003). That leaves the charge state, q, as the main factor in the mirror. The simulations assume there is no isotopic mass bias as a function of q in the solar wind.

3.2. Velocity Distribution

The overall performance of the Concentrator is affected by the long-term average energy or velocity distribution (the discussion here will use velocity rather than energy). Unfortunately, because Concentrator voltages are adjusted based on real-time proton velocity, separate runs must be made with different voltage settings for each bulk velocity modeled. Initial runs focused mostly on optimizing the Concentrator design for medium-speed solar wind conditions (e.g., 1 keV amu^{-1}, or 440 km s^{-1}), with some attention paid to extremes (300 and 800 km s^{-1}). A final set of runs was done for ^{16,18}O, He, and H, spanning the range of velocities, run at 100 km s^{-1} intervals between 350 and 750 km s^{-1}. The results of these runs were combined based on the weighting of the velocity distribution shown in Figure 2, which were taken from hourly averages in NASA's OMNI data file. Modeling of heavy ions using proton velocities ignores the observed slight velocity differences between protons and heavier ions. This velocity difference varies between ~ 0 for low-speed wind and ~ 50 km s^{-1} for high speed wind.

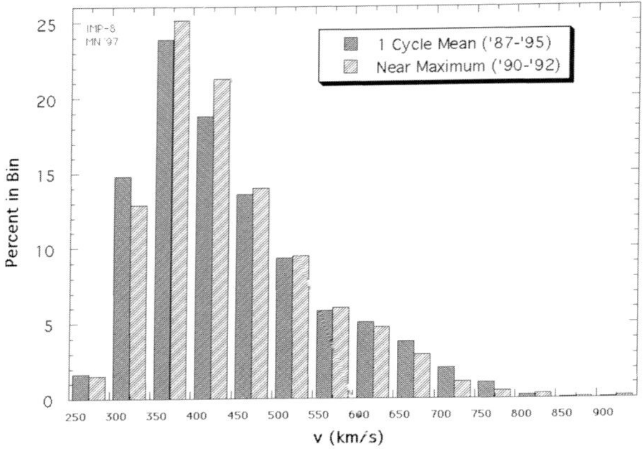

Figure 2. Proton bulk flow velocity distribution. Results were taken from the OMNI data-base of hourly averages. A companion histogram shows the distribution during an active phase of the solar cycle which roughly matches the period of Genesis observations.

3.3. ANGULAR DISTRIBUTION

The net ion angular distribution results from a number of different effects, including ion thermal distribution, bulk flow angle, spacecraft and solar wind relative velocities, spacecraft spin axis, wobble and nutation, and the instrument alignment. The overall expected angular distribution is shown in Figure 3. It is assumed that all ion species have the same thermal speed and the same distribution of bulk flow directions. Instrument alignment is within 1 deg of the spacecraft spin axis, which can be ignored. To account for spacecraft velocity around the sun, the spacecraft is pointed 4.5 deg ahead of the sun on average. Daily precession maneuvers, along with limitations on spacecraft wobble and nutation, are intended to keep the spacecraft within 2 deg of its nominal pointing. The model assumes the spacecraft is on average 2 deg off of nominal pointing. Minimizing wobble and nutation is more difficult on the Genesis spacecraft due to the movement of collector arrays. Analysis of the effect of spacecraft pointing indicates that there is very little difference in the angular distribution for higher accuracy spacecraft pointing, but the angular distribution starts to affect Concentrator performance above 2 deg off nominal pointing.

3.4. THERMAL DISTRIBUTION

Besides angular distribution, the thermal distribution causes an instantaneous spread in the velocity of incoming ions which is generally greater than the variation in bulk flow velocity over the 2.5-min duration it takes to obtain new velocity data from GIM (Barraclough *et al.*, 2003). The thermal Mach number distribution from the OMNI database for 1991–1992 is shown in Figure 4. It was mostly used to determine the efficiency of the rejection grid in getting rid of protons, based on

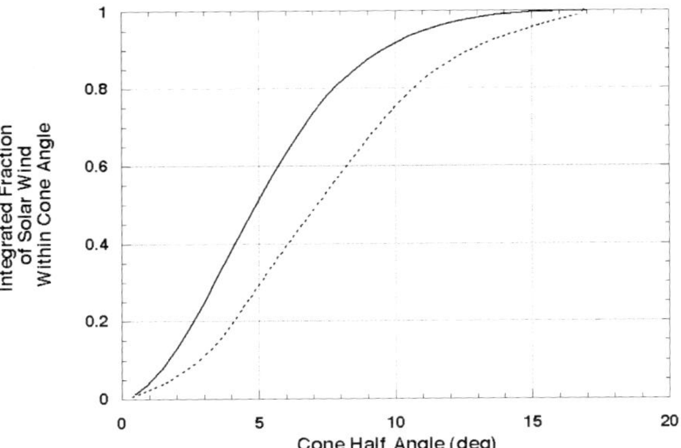

Figure 3. Calculated angular distribution of ions entering the Concentrator (solid line). The calculation takes into account thermal angles relative to bulk flow, bulk flow angle, spacecraft and solar wind vectors, and spacecraft orientation, wobble, and nutation. The curve gives the integrated fraction of the solar wind within a given cone half-angle as a long-term average. The dashed line gives the effective angular distribution for protons when grid scattering is considered for the case of 1.2 kV protons with a rejection grid voltage of 1.15 kV. Angular scattering near grid wires is an insignificant effect for heavy ions, but is clearly significant for protons.

the operation philosophy described in Nordholt *et al.* (2003). For a given thermal Mach number, e.g.,

$$M_{th} = \sqrt{(mv^2/2kT)} = \sqrt{(E_p/kT)} \tag{1}$$

and grid voltage setting, theoretical rejection fractions are relatively straightforwardly determined. Different thermal Mach numbers were tested in the Concentrator simulations in a limited way. A branch of the code was built in which the ions' parallel velocity fitted a thermal Mach number distribution, while the overall angular distribution remained the same. In these runs the overall concentration efficiency remained relatively unchanged over a large range of Mach numbers but the isotopic pattern on the target varied somewhat at low Mach numbers, as will be discussed later in Section 5.3. In reality, thermal Mach number variations will affect the angular distribution, which will in turn affect the overall collection efficiency. This level of complexity was not attempted in the model, as a mean angular distribution was used instead of additional individual runs.

4. Interactions Between Concentrator and Solar Wind

In this section we discuss physical interactions which are relevant and which were determined by means other than the SIMION model. Some of these results were

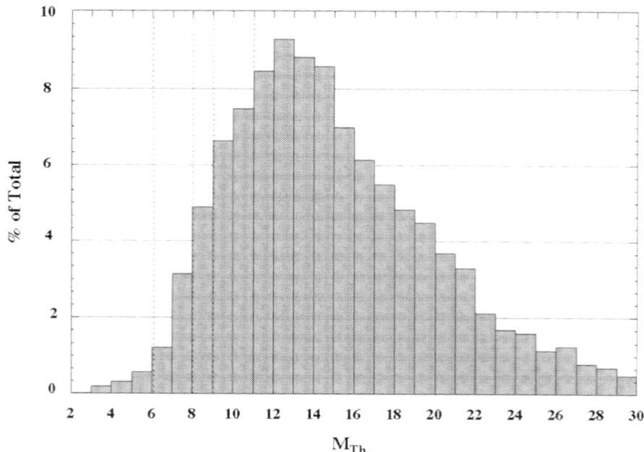

Figure 4. Histogram of hourly averages of thermal Mach numbers from the OMNI data-base for 1991–1992. Mach numbers greater than 30 are not shown. Vertical dashed lines indicate Mach values at which the Concentrator rejection grid changes voltage-to-proton-energy ratios to optimize H rejection while minimizing heavy ion mass fractionation (cf., Nordholt *et al.*, 2003).

incorporated into the SIMION model for better fidelity. This section includes a discussion of the operation of the hydrogen rejection grid, scattering of ions by individual grid wires, and backscattering of ions from the various target materials.

4.1. Proton Rejection and Heavy Ion Fractionation by the Hydrogen Rejection Grid

The Concentrator is designed to reject as much of the proton flux as possible without mass fractionating the heavier ions. If the solar wind were always a cold plasma it would be relatively simple to reject protons and collect the heavier species, as the E/q ratio, which for a given velocity is proportional to mass per charge (m/q), differs significantly between hydrogen, at $m/q = 1.0$, and all other species, at $m/q \geq 2.0$. However, as the temperature increases the E/q of the different species overlap to an increasing degree. To avoid losing heavy ions – and fractionating them in the process – some hydrogen must be allowed to pass. Figure 5 shows the fraction of hydrogen admitted through the rejection grid as a function of thermal Mach number for three different ratios (R) of rejection grid voltage (V_H) to the peak of the proton energy distribution, E_p:

$$V_H = E_p \times R. \tag{2}$$

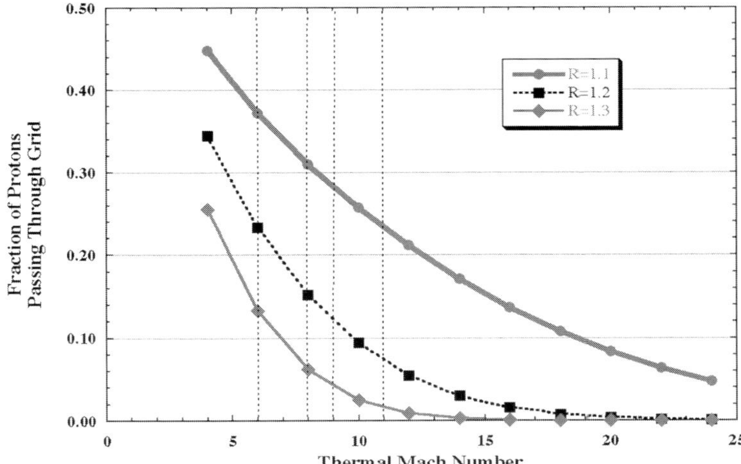

Figure 5. Fraction of hydrogen passing through the rejection grid as a function of thermal Mach number for voltage to proton-energy ratios of 1.1, 1.2, and 1.3. The Mach number data in Figure 4 combined with the results in this Figure and the voltage to proton-energy ratios given in Nordholt *et al.* (2002) Table IV predict a proton rejection of 93.8% on average.

Figure 6 shows the fractionation of $^{18}O/^{16}O$ for +8 charge species. The rejection grid operation was designed so that fractionation of this species would be under 1 permil at all times. This is done by switching R values, and hence the rejection grid voltage, as a function of thermal Mach number (cf., Table IV, Nordholt *et al.*, 2003). Because O^{+8} is usually a relatively minor charge state, except in some CME flows (T. Zurbuchen, 2002, personal communication), and is the worst case for fractionation, the average fractionation of oxygen by the H rejection grid should be < 0.1 permil.

4.2. Grid Scattering

A separate issue is the scattering of ion trajectories by passage through the grids. SIMION treats the grids as perfect membranes when in reality they are small wires with large voids between them. Electrostatic potentials poke through these voids, creating an uneven voltage 'surface'. The scattering occurs at all grids. However, the effect is greatest where the fields are the strongest on both sides of the grid, which is true of the rejection grid. The axially symmetric component of the grid scattering can be treated as a perturbation on the angular distribution of the incoming ions. When this is done, it can be shown that the difference for heavy ions is minor in comparison to the actual uncertainty of other contributions to the curve in Figure 3, such as from spacecraft pointing. But for protons, because the field perturbations near the grid wires are of the order of the proton energies near the

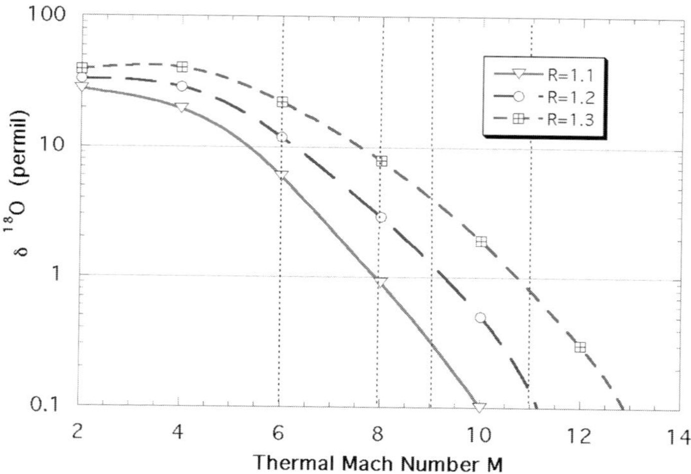

Figure 6. Calculated mass fractionation of O^{+8} by the rejection grid as a function of thermal Mach number for three different ratios of voltage to Ep used in operation. $\delta^{18}O$ is the fractionation of the $^{18}O/^{16}O$ ratio in permil, or parts per thousand].

rejection grid, the grid scattering contribution is significant. The dashed curve in Figure 3 shows the effective angular distribution of incoming protons after the grid scattering is folded in. There is significant uncertainty associated with this model due to the fact that it is a point design for one solar wind speed and Concentrator voltage setting. This suggests that the H distribution on the target is still relatively uncertain.

The non-axisymmetric grid scattering component was shown to be strongest near the center of the target, where the concentration gradient was steepest. For heavy ions it is significant only within the inner 2–3 mm. Because the target quadrants to be analyzed do not extend within 3 mm of the center, this non-axisymmetric scattering of heavy ions is of no consequence. There may, however, be measurable differences in the H fluence as a function of radial angle on the main portion of the target.

4.3. Backscattering Losses of Ions at the Target

Backscattering of ions from a surface is a well-known phenomenon. Backscattering occurs most readily for light ions incident on high atomic mass substrates, for ions with relatively low energies, and for relatively high incidence angles from the normal. The Concentrator target materials are relatively comparable in atomic mass to the oxygen ions of interest. While backscattering is relatively low for oxygen, the ions are in a region where mass is a significant factor. In other words, the mass difference between 16 and 18 amu makes a difference in the fraction backscat-

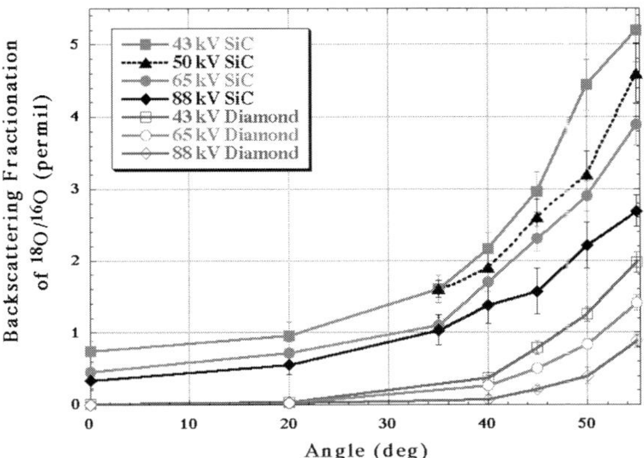

Figure 7. Backscattering mass fractionation of oxygen isotopes on SiC and diamond targets for typical Concentrator energies, as estimated by TRIM. Data points are shown with lines connecting them. The angles with respect to the target depend on the location of the ion when it exits the mirror portion of the Concentrator. See text for explanation.

tered. Additionally, the angle of incidence from the mirror to the target is relatively high, up to $\sim$60 deg. The TRIM/SRIM code (IBM) was used to estimate mass fractionation of oxygen due to backscattering as a function of target material, incidence angle, and ion energy. The Concentrator's acceleration potential ensures that ion energies are significantly higher than nominal solar wind energies. However, there is a range of energies, depending primarily on charge state. For example, a medium velocity (440 km s^{-1}) $^{16}O^{+6}$ ion will have 16 keV initial energy plus 6.5 kV $\times 6 = 39$ keV acceleration energy for a total of 55 keV. A +5 ion in the same flow will have 48.5 keV, while a +8 ion will have 68 keV. A low-speed $^{16}O^{+5}$ ion could have as little as 40 keV. The highest energy oxygen ions will be a little over 100 keV. Approximately 1 million ions were flown in TRIM for each of ^{16}O and ^{18}O at a number of energies for both diamond and SiC, the two materials used in the Concentrator target. H and He ions were also run to check their backscattering losses.

Backscattering losses are highest for H and He due to their low atomic masses, ranging up to $\sim$12% for He into SiC at relatively high incidence angles. The results for oxygen are shown in Figure 7. For each data point the ^{16}O was input with the energy given, while the ^{18}O was input with 1 keV higher energy. During periods of moderate to high solar wind velocity, the energy difference between the isotopes will be greater, but the higher energy will also mean less fractionation. The fractions of ions lost by backscattering ranged from 0.07 to 1.07% for ^{16}O into diamond and from 0.64 to 4.67% for ^{16}O into SiC over the range of angles and energies in Figure 7. To simulate backscattering in the SIMION models, the angle at which each ion struck the target was determined, and based on the angle and energy, each ion was given the appropriate probability of being backscattered.

4.4. Space Charge Effects

Because ions were flown on an individual basis, SIMION did not take into account space charge effects. Space charge effects are considered to be negligible for a number of reasons. Firstly, hydrogen, the dominant solar-wind component, is reduced by a factor of ~ 10 most of the time. Secondly, ions approach the target from a large range of angles rather than forming a collimated narrow beam. Thirdly, the ions are accelerated so that they transit any high-charge-density regions rapidly. The result is that under worst case conditions space charge effects make no more than a few nm difference in the final position of an ion on the target.

5. Ion Collection Patterns on the Target

After the design was finalized and the various aspects described above were thoroughly studied, a final set of runs was made taking into account angular distribution (with azimuthally averaged grid scattering folded in), charge state distributions, and backscattering. As mentioned above, separate runs had to be made to cover different velocity ranges, as the voltages must be adjusted between each run. Runs of up to 2 million ions were carried out for each velocity, ion, and target material, with over 20 million ions run in total. The output was a distribution of radial positions on the target for ions that were not backscattered. These were tabulated by 5 mm radial bins to allow sufficient statistics in each bin.

5.1. Concentration Factor and Collection Efficiency

The total fraction of ions hitting the target gives the average concentration once grid transparency, the geometric aperture to target ratio, and a correction for the actual aperture of the instrument are factored in. The latter factor is necessary because the radius over which ions were started, 20.8 cm, was slightly larger than the grid apertures, 20.6 cm at the ground grid and 20.0 cm at the domed grid, to allow ions to enter at an angle. Because of greater backscattering losses for the SiC portion of the target, the average concentration factor differs slightly between diamond and SiC, at 20.9 and 21.2, respectively. These concentration factors are relative to the total fluence of ions, not relative to passive collectors, for which backscatter losses would need to be considered. Concentration factors relative to passive collectors would thus be slightly higher. The fraction of ions that pass the grid structure but miss the target is relatively low, averaging slightly less than 10%.

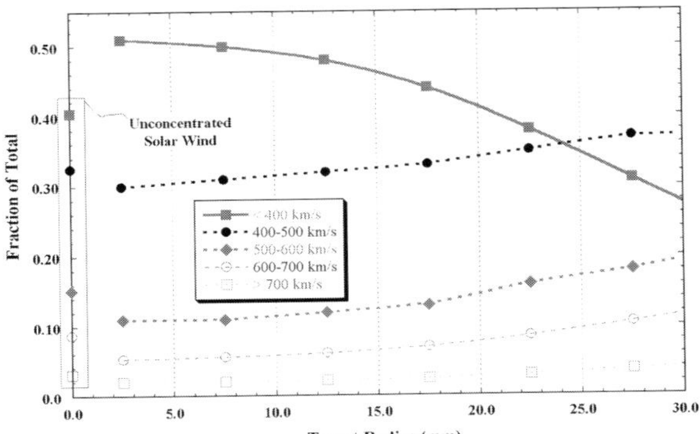

Figure 8. Relative contributions of solar wind of different velocities to the total fluence at a given target radius. The relative contributions of different velocity solar wind, from Figure 2, are shown in the box along the y-axis. The relative contributions at each 5 mm radial bin on the target are shown by data points connected by smoothed-fit curves. Low-speed solar wind represents a disproportionate fraction of the ions collected near the center, while high-speed wind is more strongly represented near the outer edge of the target.

This fraction is particularly low for low-speed solar wind, where the Concentrator is more efficient due to the increased efficiency of the acceleration voltage in straightening the incoming trajectories of lower-speed ions. For example, the fraction missing the target at 350 km s^{-1} is $< 2\%$. By 650 km s^{-1} almost 30% of the ions heading toward the target miss it. A change takes place in the Concentrator operation at 666 km s^{-1}, where the maximum mirror potential of 10 kV is reached. Consequently, ions of higher velocity reflect deeper in the mirror gap, which tends to focus them more efficiently. This plays against the decreasing effectiveness of the acceleration voltage at straightening the trajectories, resulting in approximately the same collection efficiency at 750 km s^{-1} as at 650 km s^{-1}.

A result of the more efficient focusing of low-speed ions is that the center of the target predominantly samples low-speed solar wind, while the outer portion disproportionately represents high-speed wind. This is seen in Figure 8, where the relative fractions of solar wind in a given velocity range are shown for various radial distances from the center of the target. The < 400 km s^{-1} wind has dropped at the outer edge to approximately half its relative contribution at the center. In terms of relative contribution, the high-speed solar wind increases out to the edge in Figure 8. However, in absolute terms these fluences drop from the center to the edge as well, because the total ion fluence is relatively sharply peaked at the center of the target, as discussed below.

5.2. Concentration and Ion Fluence Patterns

Figure 9 shows the concentration factor as a function of target radius for ^{16}O, $^{4}He^{+2}$, and H. The data points represent 5 mm radial averages. The curve is a smoothed fit to the data points. Error bars show the 1σ statistical uncertainties. The gradient from the center reflects, to some degree, the angular distribution of incoming solar wind ions. Ions entering at a larger angle tend to strike the target at a higher radius. However, the Concentrator does not 'image' perfectly. This is because the domed grid has significant deviations from the ideal parabolic shape built in that intentionally blur the image (Nordholt *et al.*, 2003). This blurring lowers the concentration gradient by spreading ions further from the center of the target. [As is described below and in Nordholt *et al.* (2003), this blurring is desirable because it reduces the mass fractionation gradient between the center and outer edge of the target.] The He concentration-factor curve in Figure 9 is nearly identical to that of oxygen. Helium has a mass/charge ratio of 2.0, while the majority of ^{16}O is at 2.67, with a range from 2.0 to 3.2. The similarity of these curves shows the relative insensitivity on this scale to differences in the mass/charge ratio. This is of course necessary to minimize isotopic mass fractionation, which will be discussed below. The calculated He and H concentration factors do not include backscattering losses, which would lower the expected fluences by a few percent each. The proton pattern is quite highly uncertain due to the strong interactions with the rejection grid, as mentioned in Section 4.2. The proton pattern was produced by using a thermal Mach number of 8, with the rejection grid set at a voltage to proton energy ratio, R, of 1.1 (e.g., Table IV in Nordholt *et al.*, 2003). This is the R and Mach number combination giving the largest contribution of protons, based on the Mach number frequency curves and the relative efficiency of rejection at each Mach number. The proton concentration curve was built up by the weighted contributions at various velocities, similar to the oxygen and helium calculations.

Figure 10 shows the expected ion density patterns on the target for ^{16}O, ^{4}He, and H, based on two years of operation. The mission's collection time may be slightly longer, raising all curves by up to 20%. Assuming 2 years of collection, the maximum oxygen fluence in the analyzable portion of the target (e.g., $r > 5$ mm) is 5×10^{14} cm^{-2}, or 13 ng cm^{-2}. Maximum helium and hydrogen fluences on this part of the target are 4×10^{16} and 6×10^{16} cm^{-2}, respectively, and drop off with increasing radius similar to oxygen. Radiation damage to the target from both protons and alpha particles is a potential concern. It is not possible to reject helium because its m/q ratio is in the same range as oxygen. At 4×10^{16} cm^{-2}, alpha particles are below the threshold at which damage begins to occur. Hydrogen is likely to diffuse out of SiC. As the primary radiation damage comes from blistering due to gas bubble formation within the substrate where the influx overwhelms diffusive

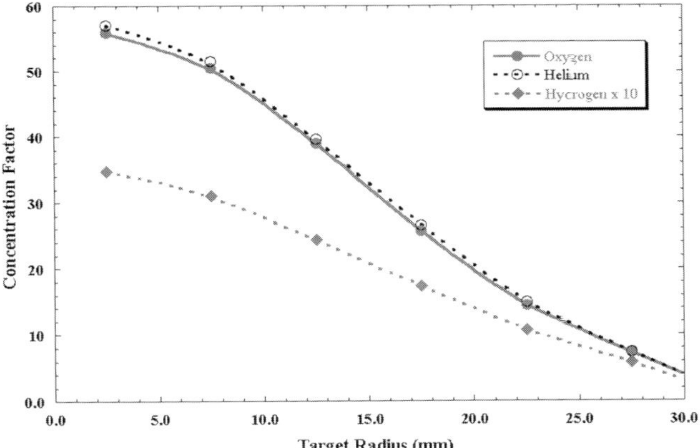

Figure 9. Concentration factor as a function of target radius for ^{16}O, ^{4}He, and H. Data points are averages for 5 mm radial bins. The curves are smoothed fits to the data points. Concentration factors for protons (magnified in this Figure by a factor of 10) are low due to the H Rejection Grid, which removes 93% of the protons to minimize radiation damage to the target.

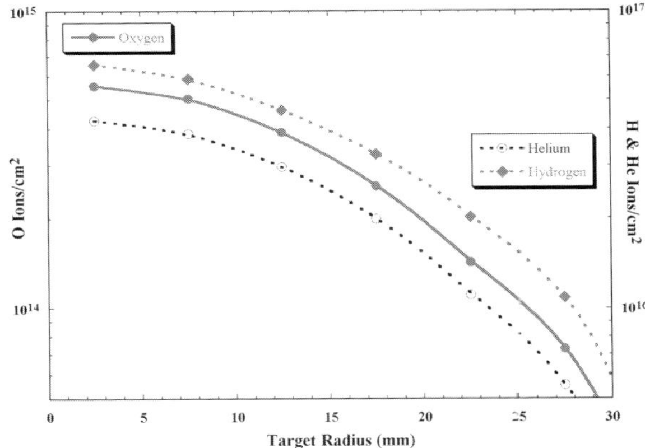

Figure 10. Predicted target ion densities for ^{16}O, ^{4}He, and H, based on a 2-year exposure time. The oxygen curve corresponds to the left-hand y-axis, while the He and H curves correspond to the right-hand y-axis.

losses, the relatively rapid diffusion of hydrogen out of SiC mostly eliminates proton radiation damage concerns there. Hydrogen is likely to be retentively held in diamond. However, the calculated proton fluence is below the threshold at which radiation damage is observed. Tests performed on diamond substrates confirmed that implanted oxygen is retentively held after a hydrogen fluence of 1×10^{17} cm^{-2} at the energies expected for ions impacting the Concentrator target.

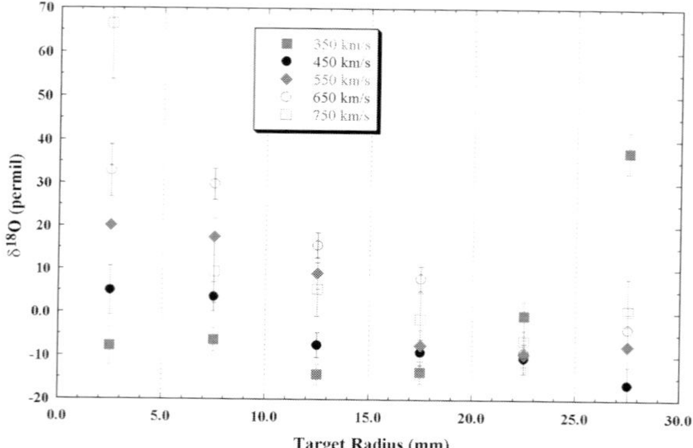

Figure 11. Instrumental fractionation for solar wind oxygen as a function of target radius for different solar wind velocity ranges for diamond target material. The design was optimized for the mean solar wind velocity of $\sim$ 440 km s^{-1}, with the result that the lower velocities have relatively flat trends.

5.3. Oxygen Isotopic Fractionation

^{16}O and ^{18}O ions were flown using the same velocity and angular distributions. The two isotopes interact differently with the Concentrator voltages because of their different energies, and they have different backscattering fractions at the target surface. The 'instrumental' fractionation for the Concentrator, averaged over the target, is -0.39 ± 0.43 permil per amu for SiC targets and -1.61 ± 0.38 for diamond, where the uncertainty is based on 1σ statistics. (Alternately, the fractionation factor, $\alpha = 0.999615 \pm 0.00043$ and 0.99839 ± 0.00038 per amu for SiC and diamond, respectively.) The difference between the two targets is solely due to backscattering. As with the overall concentration, the mass fractionation pattern varies with radius on the target, and also with velocity. In fact, because it was optimized for the mean solar wind energy of $\sim$1 keV amu^{-1} (440 km s^{-1}), the high-speed solar wind has relatively large variations in fractionation factor as a function of radius. The fractionation patterns on the target are shown for different velocity bins in Figure 11. The $<$ 400 km s^{-1} velocity bin, which comprises more than 40% of the ecliptic solar wind, is essentially flat within uncertainty over the inner four radius bins. Near the outer edge of the target, the fractionation factor rises. The acceleration grid is more efficient at focusing ^{16}O than ^{17}O or ^{18}O with their slightly higher initial energies. With a perfectly parabolic domed grid, the overall pattern would thus go from isotopically light in the center, where more ^{16}O is focused, to heavy near the edges, where the fringes of the heavy isotopes protrude past the edge of the ^{16}O pattern.

The deviations from the ideal parabolic shape in the domed grid (Nordholt *et al.*, 2003) counteract this tendency towards light isotope enrichment in the center. The light isotopes are more strongly affected by these deviations in the domed grid because, with their lower energy, they reflect closer to the grid. With a perfectly parabolic domed grid, all ions would be sharply focused near the center of the target. The domed grid deviations from a parabola serve to smear ions away from the center. This smearing effect is stronger for ^{16}O, leaving a heavy isotope enrichment near the center. This tendency is particularly obvious for higher velocity ions for two reasons. One is that the relative effect of the acceleration grid is diminished for ions with higher incoming energy. The second reason is that penetration depth into the gap between the domed grid and the mirror electrode decreases at higher solar wind energies. The closer the ions reflect to the domed grid, the greater is the effect of the domed grid deviations from parabolic. The penetration depth decreases at high velocities because the spread in penetration depth between various charge states increases.

A slight digression into the operation of the mirror is necessary to explain this point. As described in Nordholt *et al.* (2003), the mirror voltage is maintained at $4.32 \times E_p$ for all but the highest velocity solar wind. The value of 4.32 is required to keep a margin of 20% in energy on the highest expected m/q for oxygen of 18/5. That is, a $^{18}O^{+5}$ ion with energy 20% above the nominal for that proton velocity would just touch the mirror electrode if the ion were normally incident and the mirror were flat. The relative penetration depth of a normally-incident ion of mass m in amu, energy E in eV and charge q, in an ideal flat mirror is easily calculated from

$$d/D = (V_{\text{accel}} + E/q)/(V_{\text{accel}} + V_{\text{mirror}}) = (V_{\text{accel}} + mE_p/q)/(V_{\text{accel}} + 4.32E_p), \quad (3)$$

where V_{accel} and V_{mirror} are acceleration and mirror voltages, respectively. Table II gives d/D ratios for ^{16}O at different charge states and velocities for a flat mirror. The results show that the penetration depths decrease with increasing ion velocity. This occurs until the maximum mirror potential of 10 kV is reached at 666 km s^{-1}. Above this, the penetration depths again increase until the Concentrator is turned off for velocities greater than 800 km s^{-1} to prevent fractionation from occurring due to preferential loss of ^{18}O hitting the mirror. In reality, the d/D ratios in Table II are upper limits for those experienced in the Concentrator due to the curvature of the mirror and the ion angular distribution (only the energy component normal to the mirror is relevant to the penetration depth), but they show the trends with velocity.

The fractionation trends in Figure 11 show a general increase in heavy isotope enrichment near the center of the target with increasing ion velocity. This trend is consistent for data points up through 650 km s^{-1}, but the pattern changes for 750 km s^{-1} as might be expected from the fact that the mirror voltage no longer

TABLE II

Ion penetration depths d, relative to the total grid-to-electrode gap D, for ^{16}O ions of different velocities and charge states (q) normally incident on a flat (planar) ion mirror. These represent upper limits for penetration depths in the Concentrator mirror because of the mirror curvature and ion angular distribution.

q	(d/D) @ 350 km s^{-1}	(d/D) @ 450 km s^{-1}	(d/D) @ 550 km s^{-1}	(d/D) @ 650 km s^{-1}	(d/D) @ 750 km s^{-1}
+5	0.92	0.89	0.87	0.85	0.96
+6	0.89	0.84	0.80	0.77	0.87
+7	0.86	0.81	0.76	0.72	0.80
+8	0.84	0.78	0.73	0.68	0.75

matches the increased velocity. The 750 km s^{-1} data show higher enrichment within 5 mm of the center of the target (only the bottom end of the error bar is visible in the Figure), but fractionation at larger target radii is lower than the 650 km s^{-1} case. This reflects the change in interplay between the reduced acceleration grid effect and the effects of the deviations from a perfect parabolic shape of the domed grid. The relatively high fractionation at the target center for high-speed wind suggests the possibility of significantly reducing the fractionation gradient by turning the Concentrator off during high-speed streams. However, final calibration of the instrument is planned to use a comparison of neon isotopes in passive collectors with neon in the Concentrator target. If the Concentrator is not operating over the same time period as the passive collectors, this comparison will not be valid. It is not possible to close the passive collectors during high-speed streams. For this reason, the Concentrator is planned to be operated continuously during the time the passive collectors are exposed. The exception is for streams over 800 km s^{-1}, which should occur no more than $\sim 1\%$ of the time, and for which the Concentrator cannot function appropriately.

Figure 12 shows the velocity-weighted average of the fractionation curves, using the velocity distribution from Figure 2 as a weighting function. The differences between SiC and diamond are due only to differences in backscattering at the target. The fractionation gradients for both diamond and SiC targets are relatively gentle, with the curves staying close to zero permil, as expected from the fact that data points for low-speed wind, which comprise the bulk of ecliptic solar wind, were relatively flat and near unity in Figure 11. The velocity-weighted curves show heavy isotope enrichments both at the center of the target and at its edges. The center enrichment is due to contributions from high-speed solar wind, while the perimeter enrichment is due to low-speed solar wind. While the nominal values for the two target materials cross at the outer edge, they are well within uncertainty of each other. The error bars in Figure 12 reflect 1σ statistical uncertainties from the

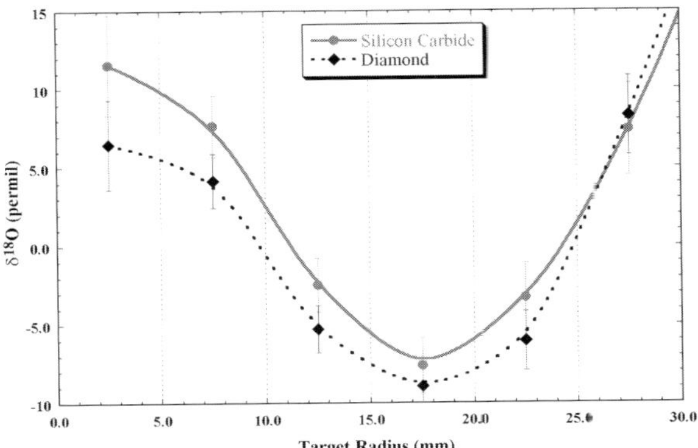

Figure 12. Calculated instrumental fractionation for solar wind oxygen as a function of target radius. This result is a weighted average of runs at different solar wind velocities (e.g., from Figure 11 for diamond). Differences between SiC and diamond target materials are due only to backscattering at the target.

numerical simulation, and do not include uncertainties such as those in the velocity distribution. Velocity distributions are known to change over the various portions of the solar cycle, and these will in fact affect the actual fractionation curves on the Concentrator target. Because the Genesis mission will be recording the velocity distribution during flight, a new weighted average based on actual flight data can be compiled upon conclusion of solar wind collection.

Besides radial distributions, azimuthal distributions were also studied to ensure uniformity as a function of azimuthal angle on the target. A series of runs at the 450 km s^{-1} setting were compiled totaling 8 million ions. Symmetries on the target were checked which corresponded to symmetry patterns in the instrument. For example, the support ribs from the H rejection grid are known to impart an azimuthal velocity component to ions approaching very close to the ribs. But while flat-grid ribs were observed to result in a 0.1% reduction in fluence to ions in the target regions corresponding to within ± 15 deg of the ribs, no fractionation was observed. Overall, symmetries corresponding to flat grid support ribs, flat grid insulators, and domed grid dimples were analyzed. A slight fractionation was observed in the case of the domed grid dimple symmetry, but the maximum fractionation was not statistically significant beyond the 1σ level (at ~ 1 permil), relative to the target-wide mean fractionation factor.

As described in Section 3.4, a branch of the ion-flying code was built which imparted thermal distributions in the parallel direction to the ions. Ions of 1 keV amu^{-1} (440 km s^{-1}) were flown with the appropriate mirror and rejection grid potentials. The results are shown in Figure 13. This modeling was done at an earlier stage, when the acceleration potential was set to 8 kV instead of the flight value of 6.5 kV. As a result, the fractionation gradients as a function of radius reported here are somewhat larger than those expected for actual operation due to the stronger

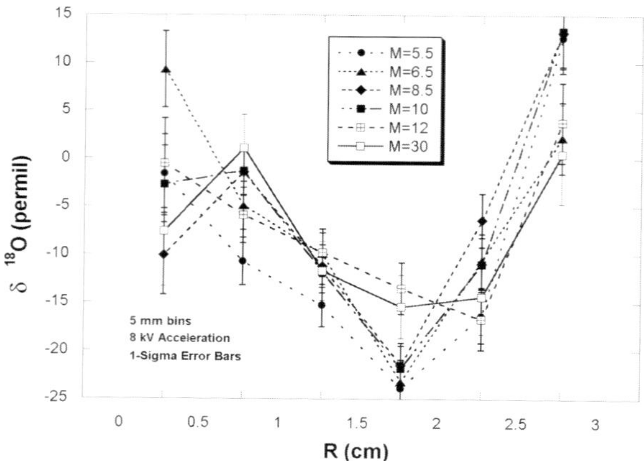

Figure 13. Instrumental fractionation patterns as a function of target radius for different thermal Mach numbers. These are for 1 keV amu^{-1} (440 km s^{-1}) ions, with energy distributions only in the parallel direction. The lines simply connect the data points to guide the eye. No significant variations are seen between Mach numbers.

acceleration grid fractionation effect. No strong differences are seen with thermal Mach numbers. There is a slight trend towards higher fractionation gradients with lower thermal Mach numbers, but it is not clear-cut. Intuitively it would seem that lower thermal Mach numbers would have smaller fractionation gradients due to the smearing implied by low thermal Mach numbers. However, at least in this simulation, the smearing is on a smaller scale than the gradients seen on the target. The actual setting of the rejection grid may play more of a role than the Mach number itself. For example, there is a big operational difference between the $M = 5.5$ data point (Rejection grid turned to 0 at $M < 6$) and the $M = 6.5$ data (R of 1.0, e.g., Rejection grid set to match the proton peak energy exactly). This may explain the significant differences at either end of the plot for these two settings. At any rate, the vast majority of the flow has thermal Mach numbers > 11 (Figure 4), for which the rejection grid voltage to E_p ratio, R, is held constant.

Because target materials will be subdivided prior to analysis, is important not only to understand the average fractionation factor for the target as a whole, but also to understand expected fractionation factors for portions of the target. This places constraints on the fractionation gradient from center to edge. For example, if cutting tolerances for sample subdividing are on the order of 100 μm, the uncertainty this produces in the predicted fractionation should still be under ± 1 permil. Looking at the curves in Figure 12, the steepest gradient is just over 1 permil mm^{-1} amu^{-1}. A cutting uncertainty of ~ 100 μm should therefore introduce not more than about a tenth of a permil uncertainty in a predicted isotopic composition for a given subdivided sample.

Another important issue is the possibility of diffusion of ions once they are collected in the target. A special effort was made to use low-diffusivity materials in the Concentrator target (cf., Jurewicz *et al.*, 2003) because of the relatively high temperatures it experiences (requirement of < 250 °C). As a result, no diffusion is expected for oxygen or helium in these materials, while diffusion of H is expected for SiC, but not for diamond.

6. Final Remarks

We have described here the performance of the unique solar-wind Concentrator with respect to oxygen isotopes, helium, and hydrogen. A great effort was made to simulate the instrument and solar-wind parameters as realistically as feasible. While these simulations were the best way of estimating the Concentrator's performance under actual solar wind conditions, the final proof of the instrumental fractionation and concentration factors is expected to be determined by comparisons of neon isotopes in the Concentrator target (and target support) with neon in passive solar wind collectors. Because of the high sensitivity of noble gas mass spectrometry, and the relative proximity to oxygen in the periodic table, neon makes an ideal comparison.

The Concentrator should be very useful for a number of other elements in the solar wind in addition to oxygen. Among these are Li, Be, B, and F which are particularly benefited due to their very low solar-wind abundances and relatively high interest for solar physics purposes (e.g., Burnett *et al.*, 2003). The Concentrator is a back-up collector for nitrogen, in that if passive collection fails to yield an accurate isotopic ratio, Concentrator targets can also be used. The utility of the Concentrator for elements heavier than neon depends on the charge state distributions of these elements. In general, elements through silicon should be efficiently concentrated, but above that, lower-charge ions will begin to be lost, culminating in an average concentration factor of less than ten for iron.

As of this writing (May 2002) the Concentrator has been in operation for nearly six months. During instrument turn-on it was discovered that the rejection grid was not able to reach full voltage for unknown reasons. A software limit was initially set at 1500 V, but it was later discovered that by limiting voltage increments, potentials of 1880 and eventually 2080 V could be reached. A software patch to limit voltage increments in the 1500+ V range was implemented to take advantage of the increased voltage capability. At nominal solar-wind velocities the rejection grid operates around 1000 V, and is not affected by this limitation. It is only in the relatively small fraction of time that high-speed streams are encountered that the voltage becomes limited by software at these levels. A calculation of the mean proton rejection over the course of the collection period yielded 86% rejection, predicting approximately twice the originally expected hydrogen fluence. One possibility is to turn the Concentrator to non-collection mode for high-speed streams. However, tests to date suggest that the increased hydrogen fluence will not result in any diffusive loss of oxygen from the target materials, though diamond targets may be near the limit of hydrogen fluence without diffusive losses. The desire to obtain concentrated and passive samples of the exact same solar wind has led the team to continue collection over all solar wind speeds < 800 km s^{-1}, as originally planned. Thus, with the exception of slightly higher proton fluences, the collection is proceeding as planned.

Acknowledgements

This work was carried out under NASA contract W-19,272. Thanks go to George Gloeckler and Thomas Zurbuchen and the respective instrument teams for making ACE and Ulysses SWICS data available. Many other people are to be thanked for their help in this work, including L. Adamic, S. Rupiper, M. Anderson, and D. Dahl.

References

Barraclough, B. L., Dors, E. E., Abeyta, R. A., Alexander, J. F., Ameduri, F. P., Baldonado, J. R., Bame S. J., Casey P. J., Dirks, G., Everett D. T., Gosling, J. T., Grace, K. M., Guerrero, D. R., Kolar, J. D., Kroesche, J., Lockhart, W., McComas, D. J., Mietz, D. E., Roese, J., Sanders, J., Steinberg, J. T., Tokar, R. L., Urdiales, C., and Wiens, R. C.: 2003, 'The Plasma Ion and Electron Instruments for the Genesis Mission', *Space Sci. Rev.*, this volume.

Burnett, D. S., Barraclough, B. L., Bennett, R., Neugebauer, M., Oldham, L. P., Sasaki, C. N., Sevilla, D., Smith, N., Stansbery, E., Sweetnam, D., and Wiens, R. C.: 2003, 'The Genesis Discovery Mission: Return of Solar Matter to Earth', *Space Sci. Rev.*, this volume.

Ghielmetti, A. G., Balsiger, H., Baenninger, R., Eberhardt, P., Geiss, J., and Young, D. T.: 1983, 'Calibration System for Satellite and Rocket-Borne Ion Mass Spectrometers in the Energy Range from 5 eV/charge to 10 keV/charge', *Rev. Sci. Instrum.* **54**, 425–436.

Geiss, J., Ogilvie, K. W., von Steiger, R., Mall, U., Gloeckler, G., Galvin, A. B., Ipavich, F., Wilken, B., and Gliem, F.: 1992, 'Ions with Low Charges in the Solar Wind as Measured by SWICS on Board Ulysses', in E. Marsch and R. Schwenn (eds.), *Solar Wind Seven* Pergamon Press, Oxford, pp. 341–348.

Jurewicz, A. J. G., Burnett, D. S., Wiens, R. C., Friedmann, T. A., Hays, C. C., Hohlfelder, R. J., Nishiizumi, K., Stone, J. A., Woolum, D. S., Becker, R., Butterworth, A. L., Campbell, A. J., Ebihara, M., Franchi, I. A., Heber, V., Hohenberg, C. M., Humayun, M., McKeegan, K. D., McNamara, K., Meshik, A., Pepin, R. O., Schlutter, D., and Wieler, R.: 2003, 'Overview of the Genesis Solar-Wind Collector Materials', *Space Sci. Rev.*, this volume.

Nahar, S. N.: 1999, 'Electron-Ion Recombination Rate Coefficients, Photoionization Cross Sections, and Ionization Fraction for Astrophysically Abundant Elements. II: Oxygen Ions', *Astrophys. J. Suppl.* **120**, 131.

Nordholt, J. E., Wiens, R. C., Abeyta, R. A., Baldonado, J. R., Burnett, D. S., Casey, P., Everett, D. T., Kroesche, J., Lockhart, W., McComas, D. J., Mietz, D. E., MacNeal, P., Mireles, V., Moses, R. W. Jr., Neugebauer, M., Poths, J., Reisenfeld, D. B., Storms, S. A., and Urdiales, C.: 2003, 'The Genesis Solar Wind Concentrator', *Space Sci. Rev.*, this volume.

Summers, H. P.: 1972, 'The Density Dependent Ionization Balance of Carbon, Oxygen, and Neon in the Solar Atmosphere', *Monthly Nottices Roy. Astron. Soc.* **158**, 255.

Wiens, R. C., Huss, G. R., and Burnett, D. S.: 1999, 'The Solar Oxygen-Isotopic Composition: Predictions and Implications for Solar Nebula Processes', *Met. Planetary Sci.* **34**, 99.

Wimmer-Schweingruber R. F., Von Steiger R., Geiss J., Gloeckler G., Ipavich F. M., and Wilken B.: 1998, 'O^{+5} in High Speed Solar Wind Streams: SWICS/Ulysses Results', *Space Sci. Rev.* **85**, 387–396.

THE PLASMA ION AND ELECTRON INSTRUMENTS FOR THE GENESIS MISSION

B. L. BARRACLOUGH[1], E. E. DORS[1], R. A. ABEYTA[1], J. F. ALEXANDER[2], F. P. AMEDURI[1], J. R. BALDONADO[1], S. J. BAME[1], P. J. CASEY[2], G. DIRKS[2], D. T. EVERETT[1], J. T. GOSLING[1], K. M. GRACE[1], D. R. GUERRERO[2], J. D. KOLAR[1], J. L. KROESCHE JR.[3,*], W. L. LOCKHART[4], D. J. MCCOMAS[1,**], D. E. MIETZ[1], J. ROESE[2], J. SANDERS[2], J. T. STEINBERG[1], R. L. TOKAR[1], C. URDIALES[2] and R. C. WIENS[1]

[1]*Los Alamos National Laboratory, MS-D466, Los Alamos, NM 87545, U.S.A.*

[2]*Southwest Research Institute, 6220 Culebra Rd., San Antonio, TX 78228, U.S.A.*

[3]*Mobius Systems Corp, 15 S. Main St. #440, Logan, UT 84321, U.S.A.*

[4]*Creative Circuitry, 1802 NE Loop 410 #15, San Antonio, TX 78217, U.S.A.*

**Presently at Conexant Systems Inc., 9020 Capital of Texas Hwy., Austin, TX 78759, U.S.A.*

***Presently at Southwest Research Institute, 6220 Culebra Rd., San Antonio, TX 78228, U.S.A.*

(Author for correspondence, e-mail: BBarraclough@lanl.gov)

Received 10 April 2002; Accepted in final form 8 August 2002

Table of contents

Space Science Reviews **105:** 627–660, 2003.

Abstract. The Genesis Ion Monitor (GIM) and the Genesis Electron Monitor (GEM) provide 3-dimensional plasma measurements of the solar wind for the Genesis mission. These measurements are used onboard to determine the type of plasma that is flowing past the spacecraft and to configure the solar wind sample collection subsystems in real-time. Both GIM and GEM employ spherical-section electrostatic analyzers followed by channel electron multiplier (CEM) arrays for detection and angle and energy/charge analysis of incident ions and electrons. GIM is of a new design specific to Genesis mission requirements whereas the GEM sensor is an almost exact copy of the plasma electron sensors currently flying on the ACE and Ulysses spacecraft, albeit with new electronics and programming. Ions are detected at forty log-spaced energy levels between ~ 1 eV and 14 keV by eight CEM detectors, while electrons with energies between ~ 1 eV and 1.4 keV are measured at twenty log-spaced energy levels using seven CEMs. The spin of the spacecraft is used to sweep the fan-shaped fields-of-view of both instruments across all areas of the sky of interest, with ion measurements being taken forty times per spin and samples of the electron population being taken twenty four times per spin. Complete ion and electron energy spectra are measured every ~ 2.5 min (four spins of the spacecraft) with adequate energy and angular resolution to determine fully 3-dimensional ion and electron distribution functions. The GIM and GEM plasma measurements are principally used to enable the operational solar wind sample collection goals of the Genesis mission but they also provide a potentially very useful data set for studies of solar wind phenomena, especially if combined with other solar wind data sets from ACE, WIND, SOHO and Ulysses for multi-spacecraft investigations.

1. Introduction

The Genesis mission is the fifth in the NASA Discovery line of competitively selected, low-cost (< \$300M) missions designed to provide frequent access to space for mid-size planetary investigations that perform focused, high-quality science. The primary science goal of the Genesis mission is to precisely determine the isotopic and elemental composition of the Sun and, by extension, the primordial solar nebula, which is the precursor material for all solar systems bodies. This is to be accomplished by exposing ultrapure collector materials to the solar wind and then returning the implanted samples from space for detailed analysis using sophisticated, ground-based instrumentation. Laboratory study of the collected solar wind samples will yield measurements of much higher quality than those currently obtainable from spacecraft-borne instrumentation. See Burnett et al. (2003) for a complete discussion of the Genesis science objectives.

The Genesis spacecraft (S/C) was launched on August 8, 2001 and is currently in a 'potato-chip' halo orbit about the L1 point, located approximately 150 million kilometers sunward of the earth, where it is collecting solar wind ions into various collector materials. The S/C is scheduled to make five, six-month orbits about L1 and then return to the vicinity of Earth in September of 2004. The Genesis Sample Return Capsule (SRC), a re-entry vehicle containing all of the solar wind samples, will separate from the spacecraft and then make a direct entry into the Earth's atmosphere above Utah. The SRC decent will be slowed by deployment of a parafoil and the vehicle will then be captured in mid-air by helicopter. Subsequently, the

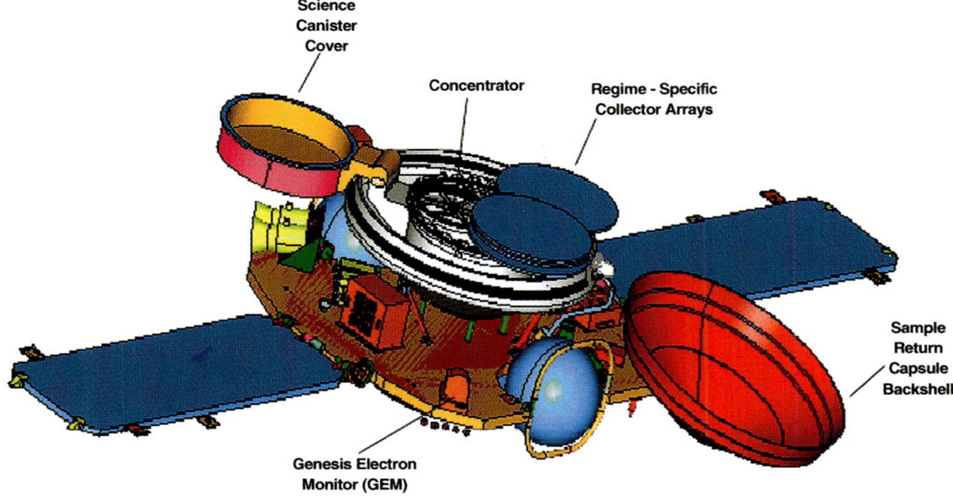

Figure 1. The Genesis spacecraft in solar wind collection configuration showing the equipment deck with attached Sample Return Capsule (SRC), solar panels and other S/C subsystems. The GIM location is directly opposite the GEM but is hidden in this view by the SRC. All sample collection materials are located in the Science Canister with some minor exceptions.

samples will be taken to clean room facilities for sample curation and eventual analysis. These will be the first samples returned to Earth from space since the last lunar samples were returned in the early 1970s during the Apollo program. A large fraction of the returned samples will be archived for study in the future when currently unknown analytical techniques and instrumentation may become available.

The entire science payload of the Genesis spacecraft consists of the plasma ion and electron spectrometers, the subject of this paper, the solar wind Concentrator (Nordholt *et al.*, 2003), which concentrates solar wind ions by a factor of ~ 20 into ultrapure collector targets (Wiens *et al.*, 2003), and various passive collector materials described in detail by Jurewicz *et al.* (2003). While not a hardware item, the WIND algorithm is also considered to be part of the science payload. This code, which resides in the S/C Command and Data Handling (C&DH) subsystem, uses the raw counts data from the plasma spectrometers to evaluate and make real-time decisions about the type of solar wind flowing past the spacecraft and to adjust the active sample collection subsystems appropriately (Neugebauer *et al.*, 2003).

The Genesis S/C consists basically of a thin, honeycomb equipment deck with the SRC, plasma instruments, solar panels and numerous other subsystems attached (Figure 1). It is spin stabilized during normal operation at 1.6 ± 0.16 rpm with the $+X$ spin axis pointing $4.5^\circ \pm 1.0^\circ$ ahead of the sun, which is the average, aberrated solar wind flow direction at L1. This orientation was dictated by the Concentrator pointing requirements, which are discussed in a companion paper (Wiens *et al.*, 2003). The electron and ion spectrometers, collectively referred to as the Monitors,

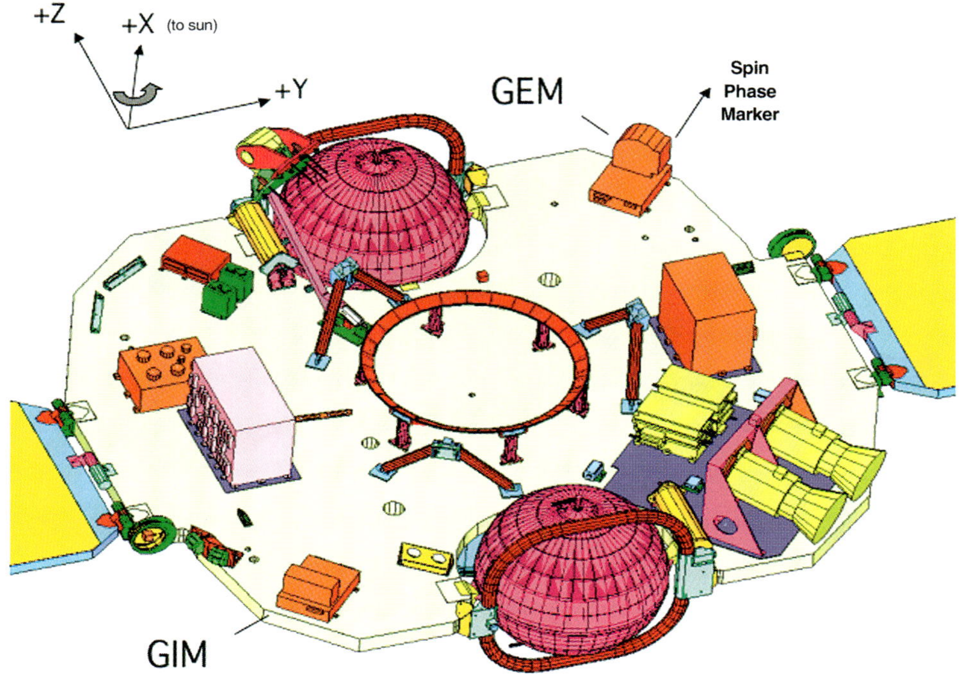

Figure 2. Layout of the Genesis equipment deck with the Sample Return Capsule removed and the locations of the GEM and GIM indicated. The GEM viewing fan is directed radially outward from the center of the deck while the GIM fan is oriented upward, with one edge of the fan lying parallel to the $+X$ axis. The center of the GEM FOV defines the current S/C spin phase angle.

are located on the S/C deck at clock angles of 45° (GEM) and 225° (GIM) relative to the $+Y$ axis (Figure 2). The S/C spins about the X-axis in a counter-clockwise direction when viewed from the Sun. Magnetic fields generated by the S/C have been controlled such that any electron entering the electron spectrometer aperture will encounter no more than a 600 nT field along its path. The use of suitable multi-layer insulation (MLI) blankets limits electrostatic potentials to < 50 mV within 0.5 m of either Monitor entrance aperture and to < 2.0 V anywhere on the S/C, except for specially waivered locations/items.

2. Requirements for the Plasma Spectrometers

Genesis is a somewhat unique mission in that the science phase does not formally begin until the flight portion of the mission is concluded and the solar wind samples have been distributed to ground-based laboratories for analysis. Despite the inclusion of plasma instrumentation in the science payload, there are no formal science goals/requirements for the Monitors as would be the case for a traditional heliospheric, magnetospheric or other space physics mission. Rather, the plasma instrumentation is present solely to support the collection of the solar wind samples and so, in this case, functions more as a S/C subsystem than as a primary science investigation. No provision was made to accommodate a magnetometer, an energetic particle investigation, a radio wave experiment, etc., as these were determined to be not strictly necessary to support sample collection. While solar wind studies are not primary science goals for Genesis, some very useful space physics can nevertheless be accomplished with the limited instrumentation available on Genesis.

The plasma spectrometers and WIND algorithm are tasked with enabling the collection of two types of solar wind samples during the course of the Genesis mission. The first type is referred to as 'bulk' sample where collector materials are continuously exposed to the solar wind without regard to the type or origin of the flow to which they are being exposed. The main operational requirement for collecting this type of sample is that the S/C pointing is controlled to within certain limits. The Concentrator, which is an active bulk-sample collector, additionally requires knowledge of the solar wind flow speed and temperature so that its collection efficiency can be continually optimized for varying solar wind conditions by adjustment of the internal ion-optics.

The second type of solar wind sample to be collected is the 'regime-specific' sample where a given Collector Array is exposed to the solar wind flow only when a specific type of solar wind is flowing past the S/C. These regime-specific solar wind samples are being collected in order to elucidate any elemental and isotopic variations that may exist in three types of solar wind flows. The three solar wind regimes are (1) the fast and fairly uniform solar wind emanating from coronal holes, (2) the slower and more variable solar wind originating in the streamer belt, and (3) the material being carried from the solar atmosphere in coronal mass ejections (CMEs). Analysis indicates that the minimum set of parameters that needs to be known onboard to reliably distinguish among the three regimes includes solar wind proton speed, temperature and density, alpha particle abundance, and the presence or absence of bi-directional electron streams. It is the measurement and interpretation of these five parameters that are the primary operational requirements for the Monitors and the WIND algorithm. The paper by Neugebauer *et al.* (2003) describes how the Monitor measurements are used to determine the solar wind regime and control the *active* collector subsystems, namely the Concentrator and the Collector Arrays, in real-time.

In contrast to many other missions, there were no particularly severe weight, power or volume constraints for the Monitors and consequently there were no extraordinary efforts taken to minimize use of these resources. Radiation hardness requirements were minimal with a total anticipated dose over the course of the mission of only 12 krad behind 60 mils of aluminum. Early in the mission design phase, there were tight telemetry constraints ($\sim$300 bps for both Monitors) and instrument operations were consequently designed to measure complete ion and electron energy spectra at the relatively slow cadence of once every 2.5 min. Much later, these requirements were relaxed considerably after a redesign of the downlink

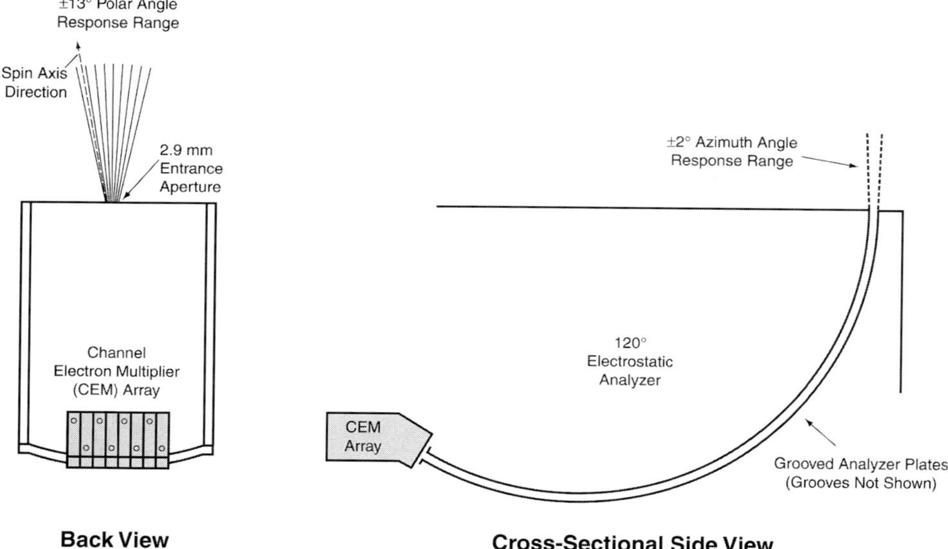

Figure 3. Simplified schematic view of the GIM sensor showing the ESA and the CEM array positioned behind the ESA exit gap. The FOV of the individual detectors is also indicated. The analyzer is mounted such that the center of the FOV of CEM #1 is aligned with the spacecraft spin axis.

strategy but by then design, fabrication and coding were too far advanced to modify the Monitor operating scheme and take advantage of the increased telemetry rate available, with the result that the relatively slow temporal resolution of the Monitors was carried forward.

3. The Genesis Ion Monitor (GIM)

3.1. GIM General Description

The GIM plasma ion spectrometer consists of a spherical-section electrostatic analyzer (ESA) for energy/charge (E/q) and angle analysis of incoming ions, followed by a custom array of Dr Sjuts Optotechnik channel electron multipliers (CEMs) for single-ion detection (Figure 3). The ESA has a 120° bending angle, a central radius of 60 mm, a nominal analyzer gap of 1.8 mm and an entrance aperture with dimensions of 1.8×2.9 mm (Table I). For UV rejection, the ESA plates have been copper coated and blackened with an Ebanol-C coating and the inner surfaces have been grooved transverse to the ion trajectories. These 0.79 mm radius grooves (0.13 mm deep with 0.84° period) have the effect of changing the effective electrostatic radii of the inner and outer analyzer plates, decreasing the analyzer constant k, defined as the accepted ion energy divided by the applied ESA voltage, from an ideal value of 16.7 to a calibrated value of 14.7. The ESA electrical configuration has the inner

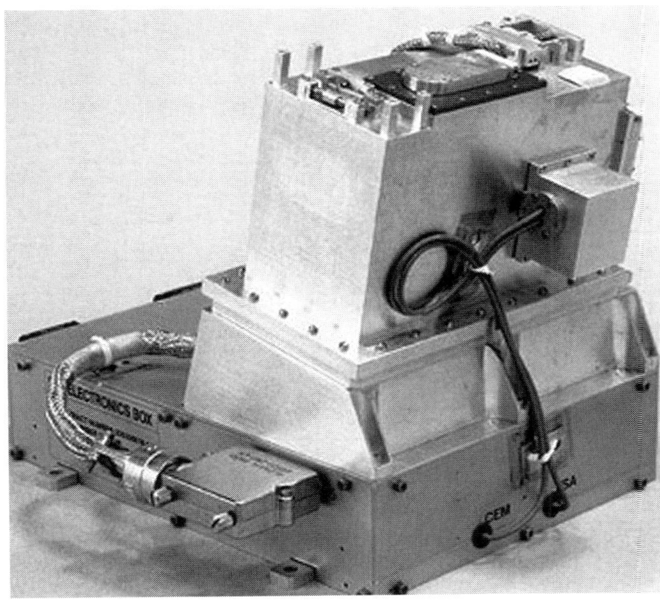

Figure 4. Photo of the GIM just prior to final preparations for S/C delivery. A wedge containing the FEE tilts the hermetically-sealed sensor 'coffin' 10.5° from the vertical to orient the FOV relative to the spacecraft spin axis. A deployable door covers the ESA entrance aperture that is located in the center of the four MLI-interface posts visible on the top of the sensor. The MEB containing the HVPS and Controller boards is the box at the bottom of the stack. Dimensions of the base are 21.6 × 16.5 cm (8.5 × 6.5 inches).

plate biased with a stepping, negative high-voltage while the outer plate is held at ground potential thus steering ions of appropriate E/q into the CEM detectors. The analyzer k and the ESA high-voltage power supply (HVPS) range define the 1 eV to 13.6 keV energy range of the sensor.

The CEM detector array consists of eight individual ceramic CEMs, each with a rectangular entrance funnel of $\sim$ 2.75 × 10.00 mm. The height of the CEMs is such that all ions exiting the curved ESA gap will be intercepted by the array, while the width and placement of each CEM was selected to obtain the desired polar angular response of the instrument. The electrical biasing scheme for the CEM array is as follows. A negative high-voltage potential (0 to −4.0 kV, −2.5 kV typical) is applied to a secondary-electron suppression grid ($\sim$90% transmission) that is mounted directly in front of the CEM funnels. This has the effect of post-accelerating analyzed ions by the amount of the applied potential. A resistor is used to drop $\sim$50 V between the screen and the CEM funnels so that secondary electrons liberated by the impacting ions are pushed back toward the funnel, thereby increasing detection efficiency. The CEM channel exit is held at $\sim$ −50 V relative to ground such that the exiting charge cloud is efficiently pulled across a small gap to the anodes that are at ground potential. The CEMs have a nominal gain of

TABLE I
GIM and GEM instrument parameters

	GEM	GIM
Species measured	Electrons	Ions
Number of CEM detectors	7	8
Energy resolution (% FWHM)	14	5.2
Azimuthal resolution (deg FWHM)	12*	4.0
(*varies with polar angle)		
Polar resolution (deg FWHM)	20	3.0
Polar FOV (deg)	150	26
Center of sensor FOV	90° from S/C spin axis	10.5° from S/C spin axis
ESA central radius (mm)	41.9	60.0
ESA bending angle (deg.)	120	120
ESA gap (mm)	3.5	1.8
k-factor (accepted energy/ESA voltage)	4.78	14.7
Aperture length along plates (mm)	10.0	2.90
Energy range (eV q^{-1})	1–1430	1–13600
Time per spectrum (min)	2.5	2.5
Mass (kg)	2.2	2.6
Power (watts)	3.67 peak 3.55 average 3.0 HVPS off	3.91 peak 3.8 average 3.0 HVPS off
Base dimensions (inches)	6.5 × 8.5	6.5 × 8.5
Telemetry rate (bps effective)	179	169

$\sim 1 \times 10^8$ and a pulse-height distribution of $< 75\%$ FWHM at 2300 V and a 3 kHz counting rate. Charge pulses collected at the signal anodes are routed to Amptek A121 hybrid preamplifiers in the front-end electronics (FEE). The discriminator threshold of these devices is controlled by a 0–5 V potential supplied by the Monitor Electronics Box (MEB). The incoming charge pulses are discriminated and amplified by the A121, and the resulting digital pulses are accumulated in 16-bit scalers until they are read out to the S/C C&DH subsystem. The preamplifiers have a fixed deadtime of 0.5 μs.

Mechanically, the ESA/CEM assembly is mounted in a 'coffin' that hermetically isolates the sensor interior from contaminants while on the ground (Figure 4). A single-use aperture door mechanism, actuated by redundant Eagle–Picher dimple motors, is used to open the aperture door in flight. Pumping of the coffin volume to space is achieved via a blackened pump-out baffle box and to a lesser extent via

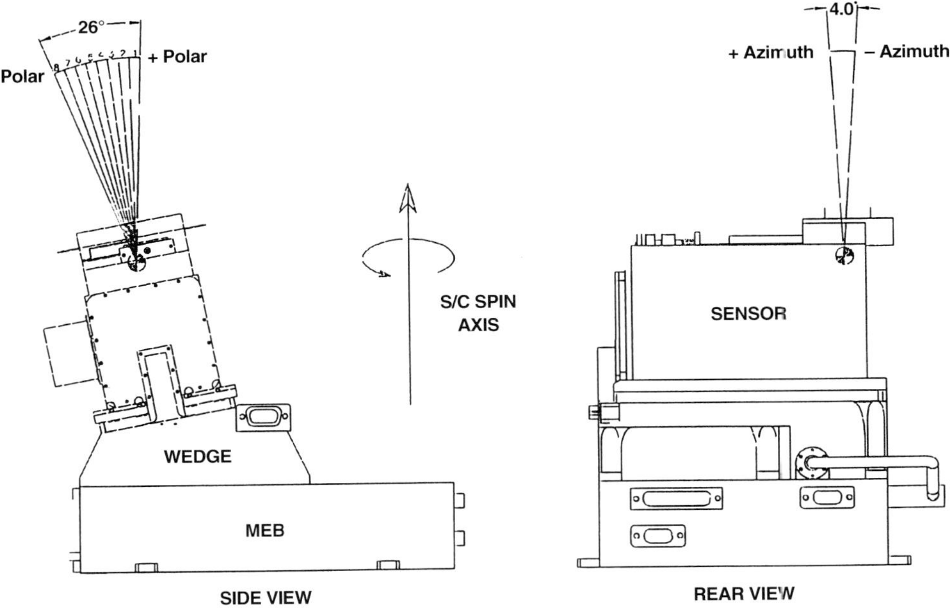

Figure 5. Drawing showing the ~ 4° × 26° GIM viewing fan and the individual CEM viewing angles. During spacecraft rotation, the fan sweeps out a circle on the sky with ~ 52° diameter.

the particle entrance aperture. As the GIM aperture stares at an angle quite close to the sun, special care was taken to baffle the area around the aperture to prevent stray light from producing any unwanted background in the detectors.

The sensor coffin is mounted on a wedge-shaped interface piece that is located between the coffin and the MEB: the MEB interfaces with the S/C deck and electrical cables. The wedge serves as a housing for the FEE and this piece is also used to position the GIM field-of-view (FOV) relative to the S/C spin axis. As each CEM has a pixel size of ~ 3° × 4° the overall GIM FOV is a narrow fan with dimensions 4° × 26° (Figure 5). The 10.5° tilt of the wedge orients GIM such that the center of the FOV of CEM #1 is parallel with the S/C spin axis (i.e., the edge of the viewing fan overlaps the spin axis by ~ 1.5°). The FOV overlap ensures that no holes will develop in the sampled phase-space as wobble and nutation of the S/C vary during the course of the mission.

The Monitor Electronics Box (MEB) is the main electronics component of both the GEM and GEM instruments and is of new design specific to the Genesis mission requirements. The unit accepts commands from the spacecraft to initiate sensor data collection and control the HVPSs. For cost-saving purposes, the MEBs for both GIM and GEM were designed to be nearly identical. The primary difference between the units is the polarity and range of the power supply outputs (Table II). (It should be noted that the redundant Concentrator Electronics Boxes (CEBs) for the Concentrator are also almost identical to the MEBs, but these are not

TABLE II
Monitor Electronics Box (MEB) output voltages

	CEM HVPS	ESA HVPS	Discriminator Threshold
GIM	0 to −4.0 kV	0 to −925 V	0 to +5 V
Accuracy (whichever greater)	± 1% or 3 V	± 1% or 0.5 V	± 1%
GEM	0 to +4.0 kV	0 to +300 V	0 to +5 V
Accuracy (whichever greater)	± 1% or 3 V	± 1% or 0.25 V	± 1%

discussed here). Each MEB has external dimensions of 5.1 × 16.5 × 21.6 cm (2 × 6.5 × 8.5 in.), weighs ∼ 1.4 kg, consumes < 4 W, and contains separate controller and HVPS printed circuit boards. The boards are partitioned for a) HVPS analog electronics (HVPS) and b) microcontroller logic/analog I/O circuits (controller).

The controller board is based on a radiation-hardened UT80C196KD microcontroller operating at 12 MHz, with the following resources:

- 16 kBytes of PROM,
- 32 kBytes of SRAM,
- 12-bit DACs (power supplies have 4096 commandable levels),
- 8-channel multiplexed 12-bit ADC,
- 1 and 100 kHz test pulser (for signal electronics chain verification),
- 16-bit scalers for FEE pulses,
- discrete inputs (enable and limit),
- discrete read backs (enable and limit),
- 19.2 kbps full duplex serial port (RS422),
- +5V and ±12 V power supply custom dc-to-dc converter,
- EMI filter and redundant power input OR diodes.

Each MEB controller contains firmware stored in radiation-hardened PROM to customize its function for either GEM or GIM. Built-in test features allow verification of ADC, DAC, counter, FEE, and spacecraft interface functions. Analog commands sent from the controller to the HVPS are looped back into the ADC multiplexer for comparison of commanded values to voltage monitor read backs.

The HVPS boards are based on a resonant flyback converter topology, operating at ∼ 100 kHz. Each power supply is scaled to accept a 0–5 V analog commands from the controller. Each output returns a 0–5 V scaled output voltage monitor that is digitized and inserted into telemetry by the controller. A safety interface that is quite useful during laboratory/spacecraft testing is provided via an external connector. Limit and enable discrete inputs are provided at this connector in parallel with the open-collector outputs from the controller board. 'Limit' sets the output voltage to $\sim 10\%$ of the commanded value. 'Enable' is used to completely turn the outputs on or off. This connector is covered with a blank for flight configuration. For safety, all power supplies are designed to operate into a short circuit indefinitely without damage.

3.2. GIM RESPONSE AND CALIBRATION

The response of the ion spectrometer was determined by illuminating the sensor entrance aperture with a uniform, monoenergetic ion beam and recording the count rate of each of the eight CEM detectors over the entire range of ESA voltage, azimuthal angle and polar angle to which the instrument is responsive. In this manner, a 3-dimensional array of transmission values, as a function of azimuth angle, polar angle and ESA potential (equivalent to incident ion E/q) was built up for each detector. The response function of the GIM can be completely determined using the resulting eight data cubes. The relative response of the GIM was calibrated to absolute beam flux using a solid-state detector with known output as a function of beam current.

Some of the results of the GIM calibration can be seen in Figures 6 and 7. The response shown in all of the 2-dimensional plots has been integrated over the energy-angle variables that are not shown. For instance, the polar vs. azimuthal response shown in Figure 7(a) has been integrated over energy, while the polar vs. transmission plot in Figure 7(b) has been integrated over energy and azimuth.

The energy response of the GIM ESA is shown in Figure 6. The transmission curves have been integrated over polar and azimuth angle and have been normalized to the maximum count rate observed for all CEMS. It can be seen that the integrated energy resolution is $\sim 5.2\%$ FWHM, which corresponds well with the 5.0% design value. The plot shows a slight variation in the central energy for the different CEMs, probably due to a small misalignment of the ESA hemispheres and/or a slight drift of the beam energy during the calibration. The integrated transmission amplitudes vary by as much as $\pm 8\%$ across the eight detectors. There is a small cosine effect that causes transmission to drop with increasing polar angle (aperture area effect) but this is not sufficient to cause all the variation observed. Probable causes of additional amplitude variation in the individual responses include variability in CEM absolute efficiency, slightly varying CEM funnel widths and small drifts in the beam current during calibration.

Figure 7(a) shows the response 'islands' of the eight-member GIM CEM array, integrated over energy and normalized to the maximum transmission for each detector. Plotted contour levels are for 90, 75, 50, and 25% response levels. The FWHM of the overall GIM viewing fan is approximately $4^\circ \times 26^\circ$ with the central response of each CEM being spaced at $\sim 3^\circ$ as desired. Figure 7(b) illustrates the GIM polar angular response integrated over energy and azimuth angle. Note that there is a very slight deviation of the response maxima from the nominal 3° CEM-to-CEM spacing due to manufacturing tolerances in the CEM widths. It can also

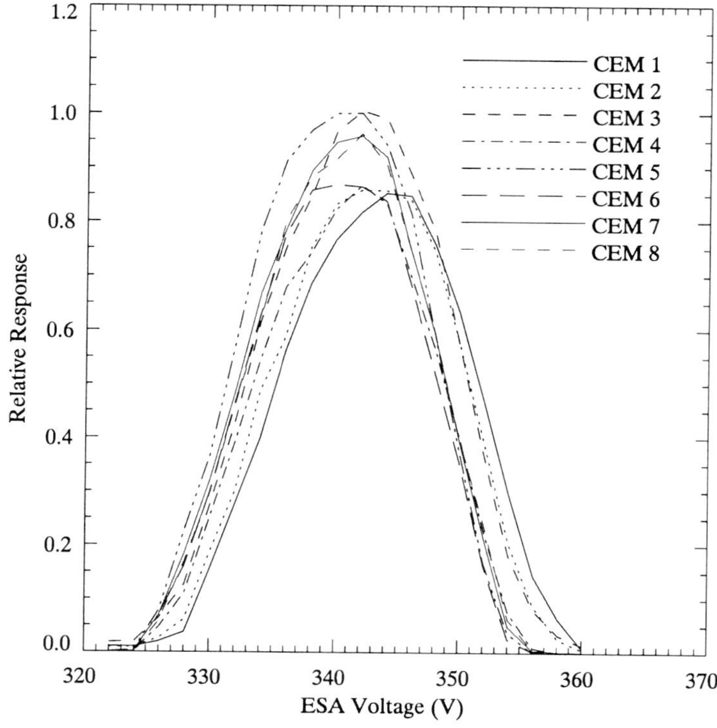

Figure 6. GIM response to a 5 keV ion beam as a function of ESA voltage (equivalent to ion E/q). Response has been integrated over polar and azimuth angle and normalized to the maximum number of counts observed. The FWHM energy response can be seen to be $\sim$5.2% and varies slightly with CEM.

be seen that there are no gaps in the polar angular coverage due to the very good overlap of the CEM responses at the 50% transmission level. The azimuthal angular response of the GIM is shown in Figure 7(c). The response curves have been integrated over energy and polar angle and the responses have been normalized to the highest transmission value observed. An integrated azimuthal resolution of $\sim$4.0% FWHM can be extracted from the curves. The $\sim 1.5°$ azimuthal offset of the central response can be attributed to the effect of fringing fields present near the entrance aperture and small mounting offsets in the calibration chamber that have not been corrected for in the plotted data.

The GIM calibration data show that the sensor meets all design goals and that there is good angular coverage of the sky with no gaps in energy or angular response. As the GIM viewing fan sweeps out a circle of $\sim 26°$ radius centered on the average, aberrated solar wind direction, GIM will adequately capture the solar wind beam distribution over the bulk of the mission.

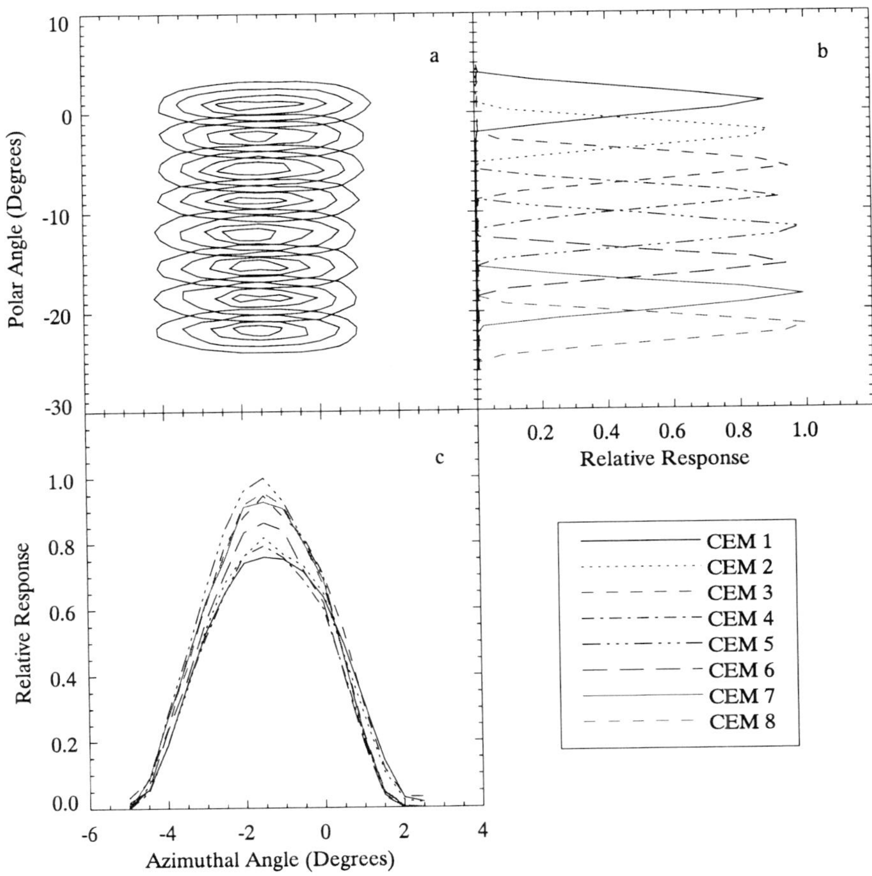

Figure 7. Calibration of GIM angular response. (a) Map of CEM response islands integrated over energy. Contour intervals represent 90, 75, 50, and 25% response levels for the individual detectors. (b) Polar response of the detectors integrated over azimuth and energy. Response curves are normalized to the maximum counts for all detectors. (c) Azimuthal response of the GIM integrated over polar angle and energy. Responses are normalized to the maximum observed.

3.3. GIM OPERATION

During normal operation, the Genesis S/C spins at a nominal rate of 1.6 ± 0.16 rpm and, due to telemetry restrictions, four spins of the S/C (nominally 2.5 min) are used to generate complete ion and electron spectra. These four spins of data collection are referred to as a complete data cycle. The operation of both Monitors is synchronized to the spin phase of the S/C and both Monitors are forced to start a data cycle and subsequently acquire data in tandem such that if the GIM is in the second of a four-spin data cycle, the GEM must also be taking data appropriate to the second spin. If, due to various problems with data acquisition, one of the Monitors calls for a repeat of a spin, the other Monitor must also perform a spin repeat (i.e., reacquire the data appropriate for a given spin). Both Monitors start their data cycles at the same time, which means that the sensor viewing fans are always 180° out of phase as the Monitors are mounted on opposite sides of the S/C (Figure 2).

The Attitude Control System (ACS) continually determines (among other things) the spin rate and spin phase of the S/C via star-tracker measurements that are acquired at a nominal rate of ~ 1.6 Hz. The S/C spin axis lies in the ecliptic plane pointing slightly to the west of the Sun. The spin phase zero azimuth for the S/C has been defined as when the center of the GEM FOV is aligned with the north ecliptic normal. At each spin phase zero crossing, ACS sends a synchronization, or 'sync' pulse to both Monitors with an accuracy of $\pm$ 10 ms. ($\sim$ 0.1 deg. of roll). As soon as the sync pulse is received, both Monitors act on a command that is already present in their command buffers and begin a new spin of operation. In the event that the ACS can't accurately determine when to send the sync pulse, a synthetic sync pulse appropriate for an exact spin rate of 1.6 rpm is continuously sent until ACS recovers. Knowledge of the S/C spin rate is also used by flight software (FSW) to continuously expand and contract (within $\pm$ 10% limits) the length of the data acquisition cycle for each spin as the spin rate of the S/C varies during the mission. This is accomplished by calculating the exact energy step-time (see below) needed for each spin so that the data acquisition cycle completely fills each spin period and all phase space samples remain aligned in look direction.

The GIM has two basic operating modes: Normal and Manual. The Normal mode has two different submodes, Search and Track, which are very similar and differ mainly in the scanned energy range. A variant of the Manual mode is the Calibrate (Cal) mode where several Manual configuration commands are sequenced to perform CEM gain measurements.

In Manual mode, a configuration command is sent to GIM, the sync pulse arrives and initiates the reconfiguration of the sensor, and the state of the sensor remains static until a subsequent command is received. This mode is useful for instrument turn-on/off, troubleshooting, periodic maintenance and CEM gain calibration. A Manual mode configuration command is used to set: fixed HV levels for the CEMs and the ESA, the discriminator voltage level, the integration, settle and step times (see below) and test pulser off/on.

GIM Calibration mode is simply a fixed sequence of Manual mode acquisition commands that is used to calibrate the gain of the CEM detectors. Basically, the ESA potential and preamplifier threshold levels are held at fixed values for the entire five-spin Cal cycle while the CEM HVPS is fixed at different values for each of the five spins. Cal mode must be entered directly from normal mode as the energy level of the proton peak from the last normal mode data cycle is needed to select the fixed ESA voltage level that will be used for the Cal cycle.

GIM Cal mode starts by determining the ESA level that gives the peak counting rate for the current solar wind conditions and fixing the ESA at this energy level for the next five spins of the S/C. The CEM HVPS is then set at a different value for

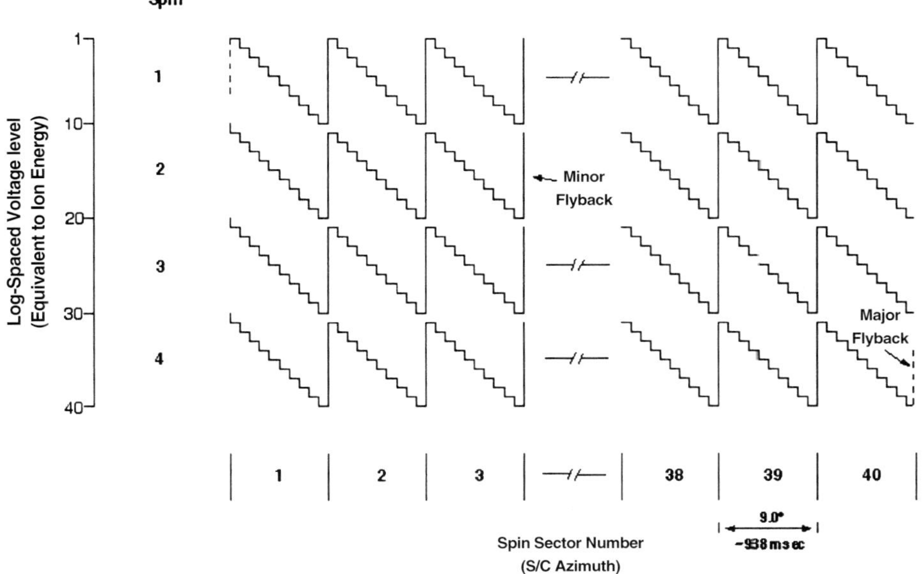

Figure 8. Schematic of a four-spin GIM Normal mode data acquisition cycle. GIM uses ten log-spaced energy steps per spin sector and forty spin sectors for a complete spin of the S/C. Four spins of the S/C are required to build a complete data cycle. The period of individual spin sector dynamically varies as the S/C spin period changes but the angular width remains constant.

each of the next five spins in the sequence -2Δ, -1Δ, $+1\Delta$, $+2\Delta$, $+0\Delta$, where Δ is a configurable number but is typically 100 V. GIM completes the Cal cycle and then returns to normal mode in its previous configuration. Data from the five CEM settings can now be evaluated to determine if the CEM gains need to be adjusted at a later date. An option exists to systematically vary the level of the discriminator threshold during a Cal cycle to determine if the current settings are appropriate. If this option is selected, the threshold voltage is stepped in the sequence -2Δ, -1Δ, $+0\Delta$, $+1\Delta$, $+2\Delta$ (Δ is configurable but typically ~ 0.3 V) at the same cadence as energy stepping occurs during normal mode operation.

As the name implies, GIM Normal mode is the operating cycle that will be used during the bulk of the mission. During standard operation, a complete ion spectrum is generated every four spins of the S/C. For each spin, GIM acquires particle counts from eight CEM detectors at ten log-spaced energies in each of forty azimuthal directions. Each of the four spins uses a different set of ten energies such that at the end of every data cycle the sensor will have acquired a data matrix consisting of detector counts from eight polar angles (set by CEM look direction), forty azimuthal angles (determined by S/C spin-phase) and forty energy levels (10 levels/spin $\times$ 4 spins/cycle).

Figure 8 shows a schematic view of a complete four-spin Normal mode data cycle. Energy stepping within a given ten-step sweep is always from high to low

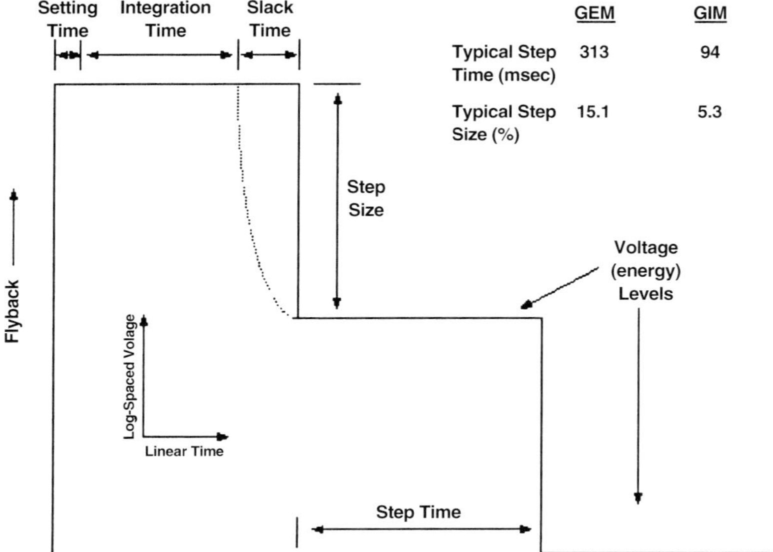

Figure 9. Schematic of the individual energy steps in a typical energy sweep. Scheme is applicable for both GIM and GEM except that different times are used for each instrument. Stepping is always from higher to lower energy levels except for flyback.

except for voltage flybacks and the ten energy levels used for a given spin are always higher than those used for the subsequent spin except when a new data cycle is initiated. A complete ten-step energy sweep takes slightly less than a second and depends on S/C spin rate.

Figure 9 shows some of the details of individual energy steps. The steps are log-spaced in voltage (ion energy) with a typical value being 5.26% spacing between steps. First, a settle time is specified that allows the HVPS to stabilize at the new voltage setting. This time is configurable over a wide range by Manual mode command but is typically held at 10 ms. When the settle time expires, integration time (also set by Manual mode command) begins and ion counts are accumulated in the scalers for the specified period, typically 40 ms. This number is decreased if counter spills are routinely encountered or increased if improved counting statistics are desired. At the expiration of the integration time, slack time begins. This period is not fixed and dynamically varies depending on S/C spin rate and specified settling and integration times such that:

$$\text{Slack time} = \text{Step time} - (\text{Settling time} + \text{Integration time}).$$

Step time is specified by FSW at the beginning of each spin as will be seen below. As soon as slack time begins, the internal microcontroller sends the ESA HVPS to the next of the ten voltage levels so that the power supply actually has slack time plus settle time to stabilize. In certain situations, slack time can decrease to zero and in this case the HVPS will at least have the fixed integration time to settle.

A Normal mode data acquisition cycle is initiated as follows. FSW is aware of the S/C spin rate and calculates the spin period for the upcoming spin. Using this information, the step time required for the 400 energy steps (40 spin sectors × 10 energy steps/sector) to completely fill the upcoming spin time is calculated. The starting level of the first voltage step is also determined (see below) as is the step size. Step size is configurable and changes depending on whether GIM is in Search or Track submode. These three parameters (step time, start level and step size) are calculated by FSW and sent to the GIM at any time between one and ten seconds before the next sync pulse is anticipated. The values are then stored in the GIM command buffer and are passed to the microcontoller as soon as the sync pulse arrives at which time the processor initiates the data acquisition sequence for the next spin.

The GIM ESA HVPS has a 12-bit (4096 level) control that can linearly step the energy acceptance of the GIM from 0 to 13.7 keV. During any given data cycle however, only forty logarithmically-spaced steps will be utilized and these steps shift in energy from data cycle to data cycle as the flow velocity of the solar wind varies with time. It is desirable to keep the solar wind proton peak centered appropriately in the energy sweep such that protons, alpha particles and higher E/q species are adequately covered even as the speed and temperature of the beam varies. This is accomplished by employing a Track submode wherein the peak count rate in the previous data cycle is used to fix the scanned energy range of the subsequent data cycle. Given a forty-level energy sweep at 5.26% spacing between levels, an attempt is made to always have the peak count rate (i.e., the proton peak) fall in level 28 (eighth step of spin #3), which will ensure that no significant portion of the proton distribution at low energies will be missed and that the high-energy tail of the alpha distribution will be adequately covered even when the beam is hot. Once the peak count rate of a given data cycle is found in energy, FSW calculates what the voltage start level of the next cycle should be to keep the peak rate in level 28. This initial data cycle start level is used by FSW to calculate the other three different start levels for each of the other spins of the complete cycle. When GIM receives the acquisition command from FSW to be used for each new spin, the internal microprocessor uses the given start level and step size to calculate the set of ten voltage steps to be used for a particular spin.

It is conceivable that the energy of the solar wind beam might shift sufficiently between the start of consecutive data cycles to cause the proton peak to fall outside of the scanned energy range. In this case, the software might lock onto the alpha peak or other feature and track this for some time, missing the proton peak completely and generating bad measurements that would confuse the WIND algorithm. To address this possibility, a Search submode has been implemented where once every 20 track cycles (the number is ground configurable), the range of the energy sweep is doubled from its normal 8× to 16× by changing the voltage step size. A fixed start level is always employed for Search, i.e., the energy range does not vary with solar wind conditions. The Search submode ensures that GIM will never track a false peak for more than about 30 min in the remote case that the proton peak jumps outside the energy range scanned in Track submode. There are also several other conditions that will invoke a Search cycle including a large jump in proton energy or insufficient flux in the proton peak.

3.4. GIM DATA AND TELEMETRY

All data acquired by GIM during a data cycle are used for onboard moments calculations while only a subset of the data selected by a masking algorithm (see below) is telemetered to the ground. Prior to being inserted into the telemetry stream, the GIM counts data are compressed from 16-bit to 8-bit numbers using an algorithm that introduces a maximum of 3% error in the transmitted counts. The combination of masking and data compression reduces the effective science data rate of the GIM from ~ 1365 bps of data sent from GIM to the C&DH subsystem to ~ 169 bps sent to the ground. GIM data are corrected onboard, prior to being used for moments calculations, for electronic deadtime and for background in the detectors. The background correction involves use of an algorithm that finds the polar and azimuthal angle of the center of the solar wind beam and then finds the azimuth that is 180° opposite. Counts data at this location for CEM #8 at the two highest energy levels are then averaged with that from the two nearest azimuthal neighbors to get a six-number background counts average. If the average is < 10 counts, no correction is made. For a background count average between 10 and 500, the correction is subtracted from each data element before further processing. If the average exceeds 500 counts (background count rate ~ 12.5 kHz) the GIM data is marked false and onboard moments processing is suspended until background rates subside.

The GIM data sampling sequence described above, in concert with the detector mounting geometry, effectively produces an over-sampling of that phase space containing the solar wind ions. In addition, the sampling sequence typically makes measurements in directions well off the solar wind beam direction. In order to reduce the amount of GIM data included in the downlinked telemetry, a masking algorithm is applied which dynamically chooses a subset of the GIM counts samples for transmission. The mask eliminates counts samples at look directions that are either highly overlapping or are at angles far away from the direction of the peak ion flux.

At each of the forty energy levels used, the GIM accumulates particle counts in all eight CEMs at forty different satellite spin azimuth positions. The samples from eight CEMs and 40 spin azimuth angles comprise a sample space of 320 phase space look directions. The masking algorithm selects 80 of those 320 look directions for downlink. As the spacecraft rotates, CEM#1 takes forty samples, all over-lapping in phase space. Samples from CEMs with FOV closer to the spin axis direction are more heavily over-sampled than those with FOV directed further from the spin axis direction. For a given data cycle, the masking algorithm takes

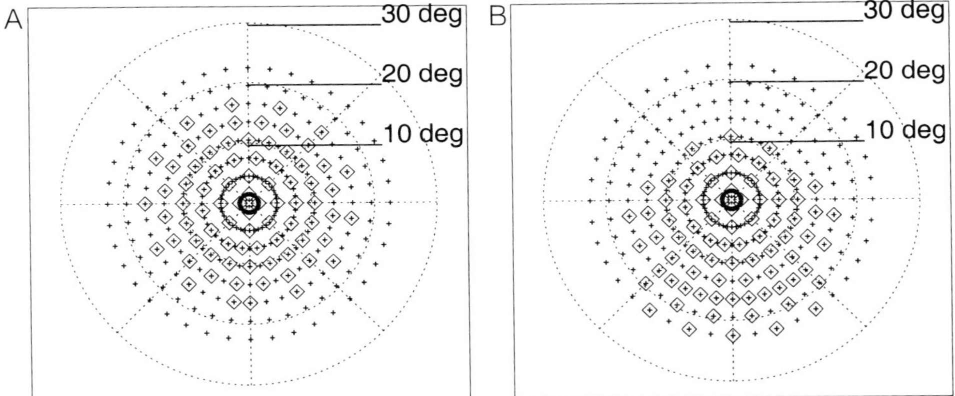

Figure 10. Mask scheme used to accommodate oversampling and reduce GIM telemetry rate. Plus symbols are samples (320 total – 40 per spin per CEM) acquired as the S/C spins and diamond symbols are the samples selected for downlink. Each of the eight rings of plus symbols represents samples obtained by an individual CEM. The selected samples should provide good phase space coverage of the solar wind beam. (a) Example of mask used when solar wind beam is located near the S/C spin axis. (b) Example of mask used when solar wind flow direction is located $\sim 6°$ off-axis.

into account the CEM and the spin azimuth angle for which the highest counts were recorded, i.e., the location of the solar wind beam. The CEM and spin azimuth angle with the maximum number of counts together define a peak sample look direction. A subset of minimally overlapping samples is chosen which are concentrated within 12° of the peak sample look direction, but have some sample coverage out to approximately 18° from the peak sample look direction.

Figures 10(a) and 10(b) show examples of what is meant by masking. On a polar projection, plus symbols indicate the look directions for the 320 samples collected for each data cycle. Diamond symbols are plotted over the pluses to show the data chosen by the mask for downlink. Figure 10(a) shows the case where the peak counts were found in the CEM closest to the spin axis. Figure 10(b) shows an example in which the peak counts were found two CEMs (i.e., $\sim 6°$) away. When flight software determines that the CEM and azimuth with the peak counts cannot be definitively chosen, GIM defaults to the mask centered on the most spin axis aligned CEM, as shown in Figure 10(a).

There are two additional data 'modes' that are available for use and these can be selected from the ground when required. These modes were primarily used for system-level testing prior to launch but will also be invaluable should problems be encountered in flight. The first mode is called 'diagnostic mode' and the main difference between this data stream and normal GIM telemetry is that the mask is not applied to the telemetered data before transmission, i.e., all 320 phase space samples are telemetered, not just the 80 normally selected by the mask. This increases the GIM telemetry rate by a factor of four from the normal rate and so must be used only when adequate onboard storage and/or downlink time is available.

Figure 11. Photo of GEM during final preparations before delivery to the S/C. The round drum is the hermetically sealed sensor head and is essentially identical to those used for ACE and Ulysses. The electron entrance aperture is located beneath the deployable door in the center of the square MLI-interface bracket on the front of the drum. The MEB base dimensions are 6.6 × 8.5 inches.

There is no comparable diagnostic telemetry mode for GEM as no masking scheme is employed with the data from this instrument (see below).

The second data mode that is available for both the GIM and GEM science data is termed ‘raw mode’. When enabled, the 16- to 8-bit compression algorithm is turned off and the full resolution scaler numbers are inserted into telemetry instead of the compressed data. This has the effect of doubling the GIM and GEM science data rates but the small errors introduced by the compression algorithm are eliminated allowing detailed analysis of unaltered counts data should the need arise.

4. The Genesis Electron Monitor (GEM)

4.1. GEM GENERAL DESCRIPTION

For the Genesis mission, it was decided to produce a copy the LANL BAM-E plasma electron experiment that is successfully being flown on the NASA/ESA Ulysses mission in order to minimize cost and reuse a proven sensor design with extensive flight heritage. The reader is referred to Bame *et al.* (1983, 1992) for detailed descriptions of the Ulysses electron sensor. The flight spare of the Ulysses instrument, with modified electronics, is also being successfully flown onboard the NASA ACE mission as the SWEPAM-E instrument (McComas *et al.*, 1998). The

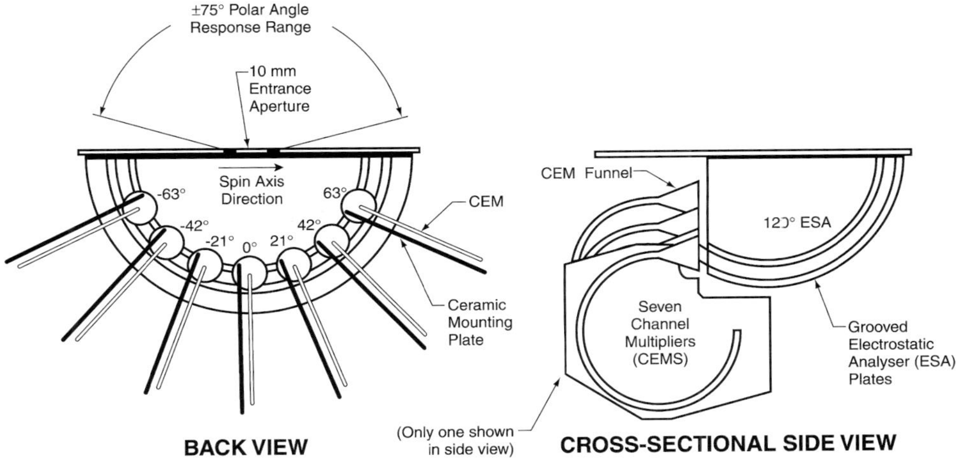

Figure 12. Simplified cut-away schematic of the GEM (after Bame *et al.*, 1983). The back and cross-sectional views show the ESA and the CEM array positioned behind the ESA exit gap. The spin axis and the viewing fan are indicated.

electron sensor (Figure 11) is somewhat oversized for the ~ 1 AU ACE and Genesis missions, as it was originally designed to make electron measurements over the 1.3–5.4 AU heliocentric range of the Ulysses orbit, but experience with the ACE instrument at L1 showed that no modifications to the design were necessary to meet Genesis performance requirements. The sensor portion of the GEM is therefore an almost identical copy of the Ulysses and ACE instruments, while the electronics are of new design and have enhanced capabilities.

The GEM sensor basically consists of an ESA, for detection and angle/energy analysis of incoming electrons, followed by a CEM detector array capable of single electron counting (Figure 12). The ESA is a spherical-section analyzer with 120° bending angle, a 41.9 mm central radius and a nominal plate spacing of 3.5 mm (Table I). The entrance aperture has dimensions of 3.5 × 10 mm. The plates are copper-coated and then blackened using the Ebanol-C process to suppress the number of unwanted UV photons that might reach the detectors and create a background. The analyzer plates are also scalloped perpendicular to the analyzed electron trajectories to further eliminate interferences from UV photons and secondary photoelectrons generated in the ESA gap. The 1.22 mm radius grooves (3° period) have an amplitude of ~ 0.7 mm and reduce the ideal analyzer k factor from 6.0 to a calibrated value of 4.78. The 0–300 V range of the ESA HVPS and the ESA k factor set the energy GEM energy range of 0 to 1434 eV.

Electrons with appropriate entry angles and energies pass through the entrance aperture and ESA, and are then detected by one of the seven Galileo CEM detectors arrayed along the ESA exit gap. Energy scanning is accomplished by holding the outer ESA plate at ground potential while a positive bias on the inner ESA plate is stepped over the desired voltage range. A high-transmission (~ 90%) grid is located between the ESA exit and the front of the 10 mm diameter CEM funnels and is biased at +200 V to provide post-acceleration to the lowest-energy electrons, thereby increasing the efficiency with which they are counted. The CEM entrance funnels are held at the screen potential while the channel exit of the CEMs can be biased from 0 to +4000 V (+2700 V typical), necessitating a capacitive coupling of the output charge pulses to the Amptek A121 hybrid preamplifiers in the front-end electronics (FEE). The charge pulses are converted to digital signals that are accumulated in 16-bit scalers in the MEB until being read out to the S/C C&DH unit. CEM gains are typically 6×10^7 with a pulse height distribution of $< 70\%$ FWHM at 2500 V and 3 kHz counting rate. The preamplifiers have a fixed deadtime of 0.5 μs and their discriminator threshold can be varied by means of the 0–5 V line from the MEB.

The polar angular response of the GEM is $\pm 75°$ with good overlap between the seven CEMs, while the azimuthal angular acceptance is $\sim \pm 6°$ (Figure 13). The $\sim 12° \times 150°$ acceptance fan is oriented such that the long dimension is parallel to the S/C spin axis and the center of the CEM #4 FOV looks along the equatorial plane of the S/C (i.e., the center of the GEM viewing fan is perpendicular to the S/C spin axis).

The GEM MEB is almost identical to that for the GIM with only a few exceptions (see Table II and MEB discussion above). The polarities of the CEM and ESA HVPSs are both positive and, due to the required resolution and dynamic range, the ESA stepping HVPS is a dual-range supply instead of the single-range type used in the GIM. Both ranges are 12-bit programmable with the low voltage range being variable from 0 to +6.39 V while the high range is commandable from 0 to +300 V. The internal microcontroller automatically handles the crossover between the ranges during energy scanning. Programming of the GEM microprocessor is also somewhat different to allow for a more relaxed data acquisition cadence and the simpler modes required for the electron sensor.

4.2. GEM Response and Calibration

The GEM response function was determined using a 2 keV ion beam (not an electron beam) and reverse biasing the ESA, i.e., negative potential was applied to the inner ESA plate to allow transmission of the ions through the analyzer. Normal biasing was used for the CEM detectors. The entrance aperture was illuminated with a uniform beam for all combinations of ESA potential, and azimuthal and polar angle to which the sensor was responsive. In this manner, a data cube of counts as a function of ESA voltage (equivalent to electron energy), polar angle and azimuthal angle was built up for each of the seven CEM detectors. The response function of the GEM can be completely determined using these seven cubes. The relative responses of the GEM were calibrated to absolute beam flux using a solid-state detector with known output as a function of beam current.

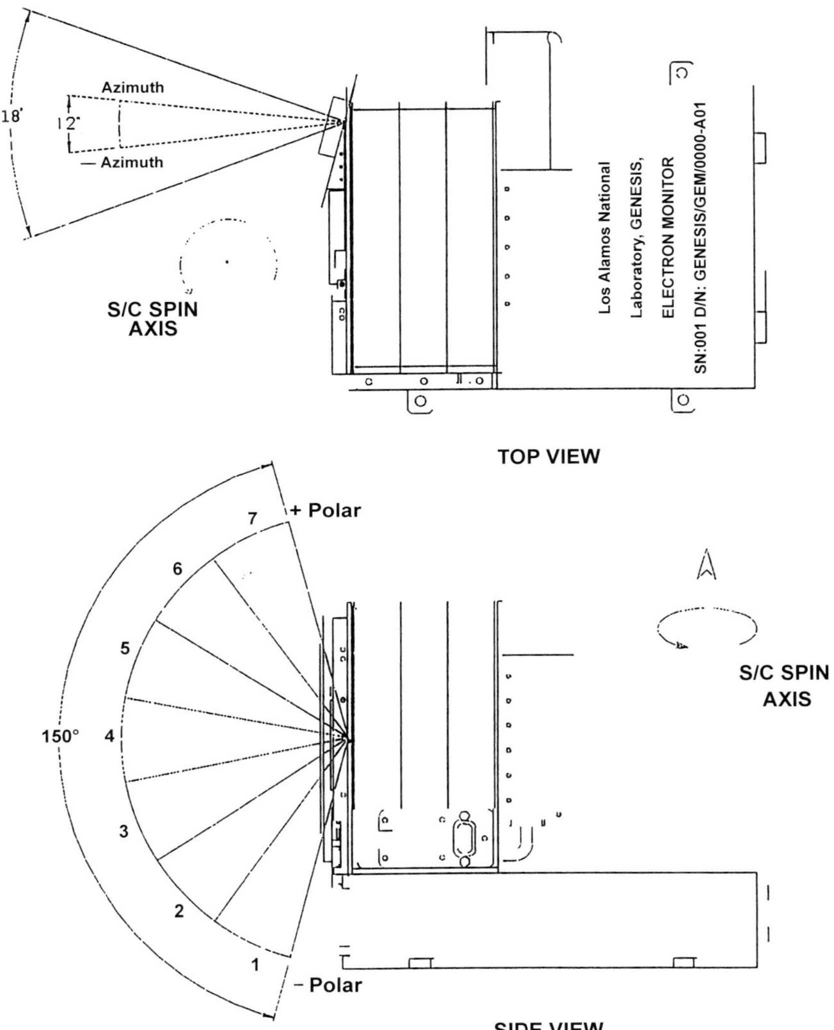

Figure 13. Mechanical drawing of the GEM showing the viewing fan orientation and the look directions for each of the CEMs. The azimuthal width of the FOV varies with polar angle when projected on the unit celestial sphere.

Figure 14 shows the energy response of the GEM ESA, integrated over azimuth and polar angle. The individual CEM traces have been normalized to the maximum transmission observed for all CEMs. It can be seen that the energy resolution of the GEM ESA is $\sim 14\%$ FWHM and that the central energy of the responses shifts to somewhat lower values as the integrated response drops (i.e., central energy decreases for the higher polar angle CEMs). The amplitude of the transmission curves is expected to vary with the polar angle due to a decrease in apparent aperture area

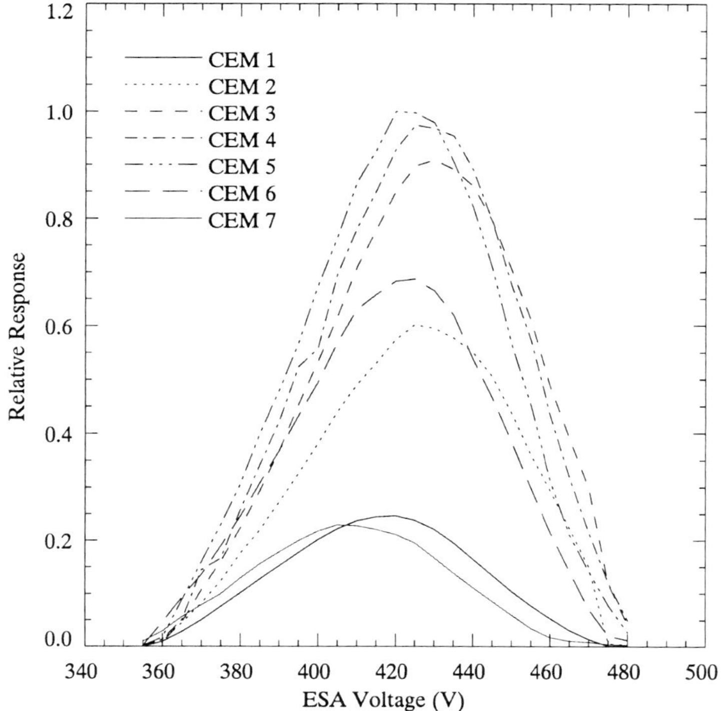

Figure 14. Energy response of the GEM ESA integrated over polar and azimuth angle. The response curves give an energy resolution of ~ 14% FWHM. A 2 keV ion beam was used for the calibration.

with increasing polar angle. Some variation in the height of the transmission curves may also be due to variation of the individual CEM efficiencies and possible slight drifts in the beam energy during calibration.

Figure 15 contains three plots of the GEM angular response. Figure 15(a) is a plot of the 'islands' of angular response integrated over energy, with each CEM island normalized to the maximum counts for that detector. The contours shown are for 90, 75, 50, and 25% response levels. It can be seen that there is good coverage of space by the $\sim 12^\circ \times 150^\circ$ viewing fan with adequate detector overlap. A broadening of the azimuthal response at high polar angles is only somewhat visible in this plot, which uses the system of coordinates in which the data were acquired in the laboratory using an azimuth-elevation setup similar to an alt-azimuth telescope mounting. The azimuthal response increase with polar angle is a natural effect that is inherent in spherical-section analyzers with 'long' entrance apertures and also arises from the practical necessity of having to offset the aperture from the biased ESA plate to avoid electrical shorting (Gosling *et al.*, 1978). The broadening effect is real and is much more apparent when the response is plotted in spherical-polar coordinates. The maximum response in the center of each of the islands can also be

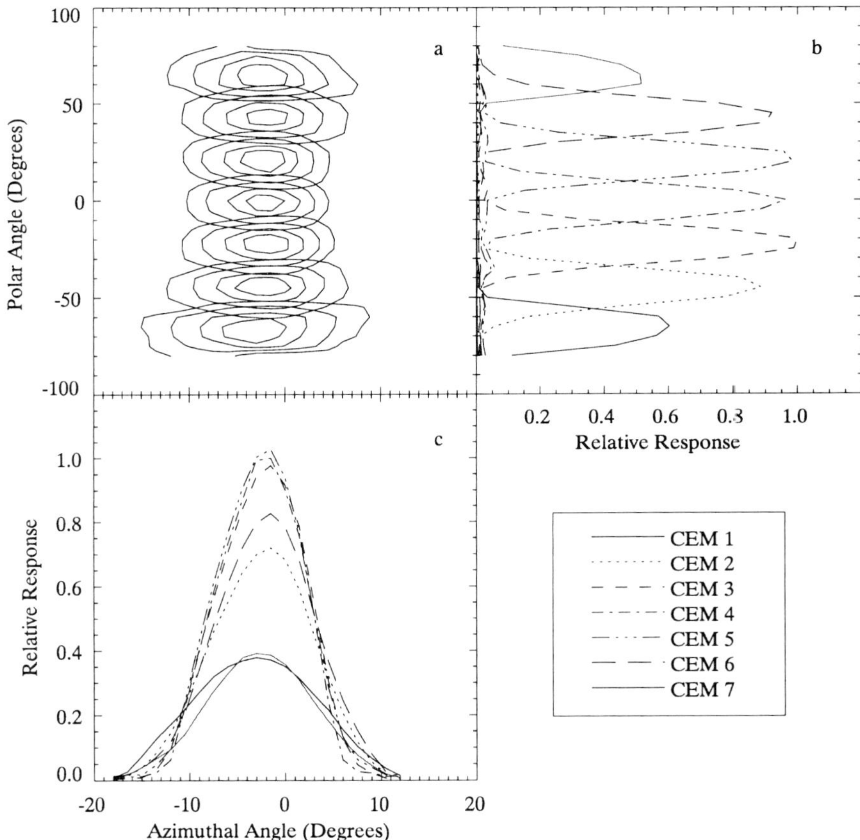

Figure 15. Calibrated angular response of the GEM. (a) Map of the transmission islands integrated over energy. The contours show 90, 75, 50, and 25% transmission levels for each detector. (b) Polar angular response for each of the detectors integrated over azimuth and energy. The response curves have been normalized to the maximum transmission for all detectors. (c) Azimuthal angular response integrated over polar angle and energy. The detector transmissions have been normalized to the maximum observed for all detectors.

seen to describe a slight curve with polar angle, with highest polar angle detectors having the central response at more negative azimuths, as expected.

Figure 15(b) shows the GEM polar acceptance, integrated over energy and azimuth. The transmission values for the CEMs are normalized to the highest value observed for all detectors. The good overlap between the detectors at the $\sim 50\%$ level is apparent, as is the drop-off in transmission at high polar angles due to a cosine effect causing a decrease in apparent aperture area with increasing polar displacement. Polar angular resolution can be seen to be $\sim 21°$ FWHM, in good agreement with the calibration results of the Ulysses and ACE electron instruments (Bame *et al.*, 1986, 1992; McComas *et al.*, 1998). Transmission centers are within $\pm 1°$ of their design values.

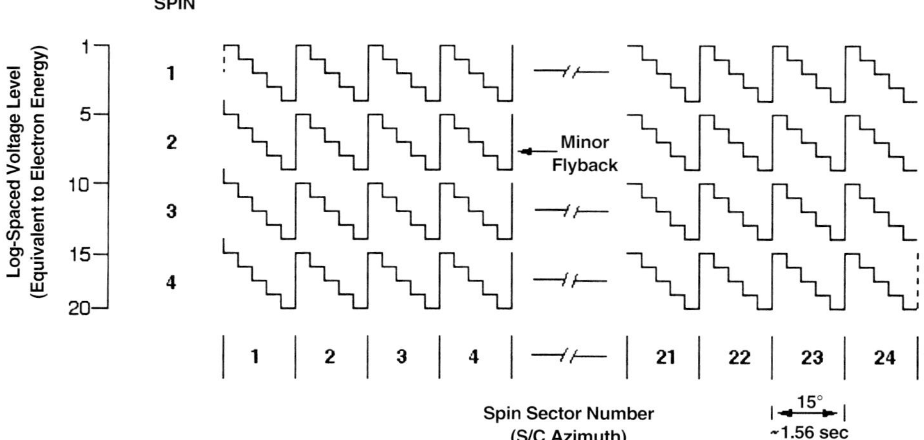

Figure 16. Schematic of a four-spin GEM Normal mode data acquisition cycle showing the twenty, log-spaced energy levels and twenty four azimuthal spin sectors used to sample phase space during a complete data cycle. The period of the energy steps expands and contracts as S/C spin rate varies.

The transmission of the GEM ESA is shown in Figure 15(c) as a function of azimuthal angle, and is integrated over energy and polar angle. The individual response curves are all normalized to the highest value observed for all detectors. Here the transmission drop-off as a function of polar angle is more readily apparent (due to the projection) and the slight variation in central acceptance with polar angle is also visible. The integrated azimuthal resolution of the GEM can be seen to be $\sim 12°$ FWHM for the detectors at the lowest polar angles broadening to $\sim 18°$ the highest.

4.3. GEM OPERATION

The operation of the GEM is similar to but somewhat less elaborate than that of the GIM. The basic Normal, Manual and Cal modes are almost identical to those discussed above for GIM but, as GEM continually scans a fixed, configurable energy range when in Normal mode, there is no need for the search cycles or the proton peak tracking employed by GIM. Initiation of data acquisition is identical to that for GIM: the S/C sends a Normal mode acquisition command followed by a sync command to start data acquisition for each spin of the S/C.

Four spins of the S/C (~ 2.5 min) are used by GEM to acquire a 3-dimensional measurement of the plasma electron distribution over $\sim 96\%$ of the 4π steradian unit sphere. A complete distribution measurement consists of counts collected from seven CEM detectors (polar angle) at twenty log-spaced ESA levels (energy) across twenty-four spin sectors (azimuthal angle). ESA voltage stepping is always from high energies to low energies. During the first spin of the four-spin data cycle, the highest five energies are scanned twenty four times, the second set of five energies are continuously scanned during the second spin, etc., until the data cycle is complete at the end of the fourth spin (Figure 16). The period of each energy step in an energy scan is automatically adjusted to compensate for any variations in S/C spin rate. Nominal energy step times are 0.313 s but these can vary by $\pm$ 10% as spin rate increases or decreases during normal operations. Count integration times can be varied over a wide range to avoid counter spills or to improve counting statistics. See the discussion in Section 3.3 for details of the energy step timing.

One of the main differences between GIM and GEM energy scanning is that the range covered by GEM is selected by command from the ground and remains fixed until the configuration is next changed. The measured energy range is easily configurable and can be set to cover any portion or all of the interval between 0 and $\sim$ 1.4 keV, albeit with varying overlap of the energy response functions at the adjacent energy levels. The highest energy desired is selected and a step interval is set: this determines the amount in percent that each subsequent energy step should lie below the previous level. Typical values used are 287 V and 15.1%, which gives a scanned energy range of 61 to 1372 eV with good overlap of the energy response between levels. This energy range avoids the photoelectron distribution and gives good coverage of the interval where counter-streaming electron distributions are typically observed.

4.4. GEM DATA AND TELEMETRY

As opposed to the case for GIM, all of the GEM counts data are sent to the ground for analysis, i.e., there is no masking of the GEM data as is the case for the GIM (see masking discussion above). The 3360 16-bit scaler values for a complete data cycle are all read out from the GEM MEB to the C&DH unit, compressed to 8-bit numbers and inserted into the telemetry stream. The compression algorithm introduces a maximum error of $\sim$3% in the counts data. The effective science data rate of the GEM is reduced from $\sim$358 bps to $\sim$179 bps (only marginally higher than the 169 bps GIM data rate) by use of the compression scheme. The uncompressed data is used on-board by the WIND algorithm to determine the presence of bi-directional electron streams (BDES) but is first corrected for electronic deadtime and detector background. The background correction involves an algorithm that averages the counts data from all seven CEMs at the highest energy level and at all twenty four azimuths to arrive at an average background counts correction. If the average is $<$ 10 counts, no correction is made. For a background count average between 10 and 500, the correction is subtracted from each data element before further processing. If the average exceeds 500 counts (background count rate $\sim$2.5 kHz) the GEM data are marked false and BDES processing is suspended until background rates subside.

5. Monitor Simulators and System-Level Testing

Realistic end-to-end testing of integrated payload components is an important aspect of any S/C test plan. This is even more important for Genesis as GIM and GEM provide raw plasma counts data to the S/C C&DH subsystem, which then processes the data into moments and parameters, makes solar wind regime determinations, and issues commands to the Concentrator and Collector Arrays, all via the WIND algorithm. But complete end-to-end testing was not strictly possible due to an inability to realistically simulate the plasma environment at the S/C testing level. To still allow nearly end-to-end testing, a system was devised in which simulated raw counts data from GIM and GEM were generated and injected into the S/C C&DH for testing of the remainder of the science payload, including autonomous commanding of the collector arrays when the simulated data indicated a change in solar-wind regime and the spacecraft ability to telemeter these regime changes and array motions to a ground station. The only item that was not truly end-to-end was the fact that GIM and GEM had to be tested separately with ion and electron beams in a vacuum chamber.

The near end-to-end simulation tests followed a several-step process as follows. Solar-wind ion data recorded for various time periods by the ACE SWEPAM-I instrument, often modified to test a particular portion of the WIND algorithm, were used for the simulation tests. The data were input as ground-processed moments into the GENSIM code that takes ACE moments and converts these data into raw counts data anticipated for GIM, reversing the process that normally produces moments from raw counts. For the GEM data, a simple bi-directional-electron index was assigned for each data cycle. These simulated ion and electron counts were then input from PCs into GIM and GEM electronic simulator boxes. The simulator boxes look identical to the actual GIM and GEM instruments to the spacecraft from an electronic and signal-processing standpoint. The simulator boxes were used both for initial safe-to-mate tests (they were available for these tests long before the actual instruments were ready) and for the software testing. The simulator boxes produced data packets simulating in-flight GIM and GEM data, which were then fed to the spacecraft C&DH. The C&DH in turn processed the data packets using the WIND algorithm described in Neugebauer *et al.* (2003) and made solar-wind regime selections accordingly.

The simulation scheme was first used to extensively test the WIND algorithm in the Software Test Laboratory (STL), which simulated the S/C C&DH and its environment. A number of tests were devised, as listed in Table III, to test various solar-wind regime transitions, and various other software features such as shock detection, despiking of noisy data, dropout of various types of data under abnormal solar-wind conditions or S/C operation, and turning the concentrator voltages to standby for excessively high wind speeds.

The STL tests were performed over a period of nearly one year prior to Genesis launch. Overall they were very useful in debugging the Moments Extractor Code,

TABLE III

Solar wind simulation tests used to verify correct onboard processing of the raw GIM/GEM data and correct performance of the WIND algorithm (see Neugebauer *et al.*, 2002).

Test	Description
F1	Despiking of data
F2	Test for proper performance during alpha data drop-out
T1	Fast to slow wind regime transition
T2	Fast to CME wind regime transition
T3	Slow to fast wind regime transition
T4	Slow to CME wind regime transition
T5	CME to fast wind regime transition and concentrator-to-standby test
T6	CME to slow wind regime transition
24 hr test	Long-duration run to test features having longer time constants
50 hr test	Long-duration run performed in STL shortly before launch

which processes the raw data to useful parameters, and the WIND algorithm, which selects the solar-wind regime and issues commands to the Concentrator and Collector Arrays. Only one of the listed tests was performed on the spacecraft, on two different occasions, to test the near-end-to-end performance, including moving the collector arrays. Performance in flight proved the success of these tests, as only one surprise was found in the code in flight, and that was in the BDEs section, which was not particularly well tested by this scheme (Neugebauer *et al.*, 2003).

6. Initial Results

GEM began making solar wind measurements on August 23, 2001 and GIM was turned on for the first time the following day. As of June 2002, only a few days of measurement time have been lost due to S/C safe-mode entry, trajectory correction maneuvers, S/C reconfiguration, etc. and no anomalies have been noted in the instruments or their operation. The level of solar activity since turn-on has been relatively high with the S/C having already encountered a number of solar energetic particle events, some with very high particle flux levels. As expected, instrument backgrounds generated by penetrating radiation have been considerably higher in GEM than in GIM as (1) the detector biasing in GEM attracts the numerous secondary electrons produced by energetic particles in the sensor interior and (2) the GEM wall thicknesses are considerably less than for GIM, thereby allowing easier penetration of energetic species. No unwanted backgrounds from direct solar UV

Figure 17. A polar plot centered on the average solar wind direction at L1 using Diagnostic mode data and demonstrating angular resolution of the GIM. The plot can be thought of as a view of the sky centered $\sim 4.5°$ ahead of the sun with radius of $\sim 26°$. Each ring defines the forty samples per spin obtained by a given CEM. The degree of sample overlap can be seen to increase in the more sunward detectors. The polar angular width of each ring is $\sim 3°$. The solar wind beam can be seen to be displaced slightly to the upper right of the diagram as evidenced by the counts modulations seen most prominently in CEM #2–4.

light leaks, or from internal UV-generated secondary electrons, have been observed in either GIM or GEM while the S/C is in its normal orientation.

Figure 17 shows a color-coded, polar plot of GIM counts obtained shortly after instrument turn-on when Diagnostic mode data was available. Each of the eight rings represents one of the CEM detectors with the central disk representing CEM #1, which is looking along the S/C spin axis toward the nominal solar wind direction. Clock angle around the plot represents azimuthal look direction (there are forty azimuthal samples around each ring) and increasing radial distance from the center is equivalent to increasing polar angle from the spin axis. Each ring is $\sim 3°$ in width. The out-of-the-page normal to the center of the plot points in the direction of the Sun and lies in the ecliptic plane. The area shown can be thought of as a

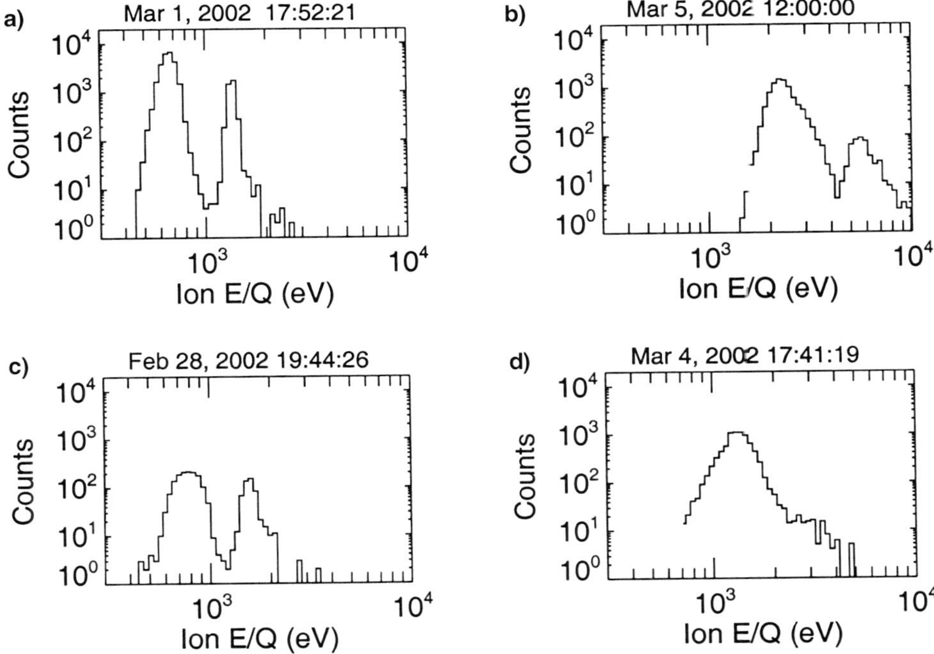

Figure 18. Four ion energy vs. counts spectra from GIM for four separate time periods illustrating the energy resolution of the instrument. Spectra are constructed from counts obtained by the single detector with the maximum counts and are integrated over azimuth. See text for an explanation of the solar wind conditions extant for each spectrum.

circular FOV of the sky with half-angle of $\sim 26°$ centered $\sim 4.5°$ west of the center of the Sun. Counts have been integrated over all energies.

It can be seen that the solar wind is fairly well centered in CEM #1 and that the beam is rather cool, as counts drop off very sharply with polar angle from the center of the FOV. There is a slight displacement of the flow toward the upper right of the plot as is evidenced by a small modulation in the counts most easily seen in CEMs #3 and #4. The high degree of oversampling in the center of the plot can easily be seen and serves to illustrate why the data masking scheme can be employed to greatly reduce telemetry while having little effect on angular resolution. Subsequent analysis of the solar wind data indicates that the flow speed during this period was ~ 360 km s^{-1}, density was ~ 0.9 cm^{-3}, and proton temperature was $\sim 8.9 \times 10^4$ K.

To illustrate the energy resolution of the GIM, four energy spectra are plotted in Figure 18. Each spectrum was acquired over one 2.5-min data cycle, is constructed of 40 ms count samples obtained by the detector with the maximum counts at forty different energies, and is integrated over azimuthal angle. Figure 18(a) shows a slow solar wind flow with a typical proton temperature ($V_p = 361$ km s^{-1}, $T_p = 6.0 \times 10^4$ K, [He] $= 6.2\%$), Figure 18(b) is a fast stream with a typical temperature ($V_p = 679$ km s^{-1}, $T_p = 3.4 \times 10^5$ K, [He] = 4.7%), Figure 18(c) shows a CME flow with low temperature and a relatively high alpha abundance ($V_p = 389$ km s^{-1}, $T_p = 2.8 \times 10^4$ K, $\rho_p = 6.4$ cm^{-3}, [He] = 8.6%) and Figure 18(d) is compressed solar wind at the leading edge of a solar wind stream that has higher than typical temperature ($V_p = 520$ km s^{-1}, $T_p = 3.9 \times 10^5$ K, [He] could not be determined accurately). These figures serve to illustrate that the track energy range of the GIM is adequate to cover the proton and alpha peaks for almost any solar wind condition that will be encountered, that the very good energy resolution can easily resolve the proton and alpha peaks except at the hottest temperatures where the distributions become inextricably overlapped in E/q, and that the peak resolution is more than adequate for determining good plasma moments. The reader is referred to the companion paper by Neugebauer *et al.* (2003), which gives the details of how the GIM counts spectra such as these are converted on-board to plasma moments and how these moments are used to make real-time adjustments of the collector subsystems. Several examples of moments time-histories derived from GIM flight data are also presented there.

Figure 19 illustrates the ability of the GEM to meet its requirements of being able to determine the presence/absence of bi-directional electron streams. This parameter is one of those used in the real-time identification of CME flow past the S/C. Figure 19(a) is a series of nine energy-cuts (of twenty available onboard) showing the location in polar and azimuth angle of the solar wind electron strahl. Note that only a single beam is evident, centered at $\sim 70°$ polar and $100°$ azimuth angle, but it is clearly visible over the range of at least 200 to 1000 eV.

Figure 19(b) is an identical series of nine energy-cuts taken ~ 25 hr after the previous example but here a counter-streaming beam, located $\sim 180°$ opposite the antisunward-strahl, is now clearly visible. Such bi-directional electron streams can sometimes be encountered in CMEs when regions with closed magnetic field lines transit the S/C. The WIND algorithm checks that at least three energy levels in the appropriate energy range contain significant evidence of counter-streaming electrons before declaring their presence valid.

7. Conclusions

The GIM and GEM are currently providing fully 3-dimensional measurements of the solar wind ion and electron populations at L1 in support of the primary operational goal of the Genesis mission, namely, the collection of ultrapure solar wind samples for return to Earth. These measurements are currently being successfully utilized onboard to autonomously control the Concentrator and the Collector Arrays. Results to date show that we are able to derive high quality solar wind parameters from GIM and GEM. These parameters will be used to produce a record of the Genesis sample collection history specifying the solar wind particle fluence and probable solar wind flow type as a function of time during the

A. GEM 2001 OCT 21 11:21:38

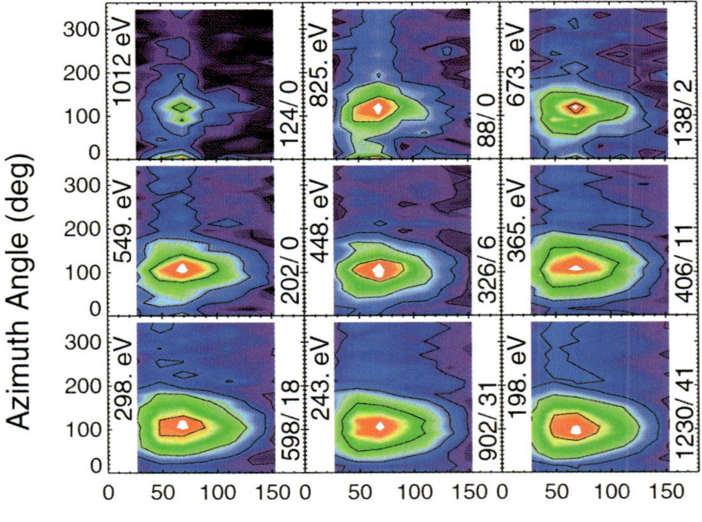

Polar Angle (deg)

B. GEM 2001 OCT 22 12:47:36

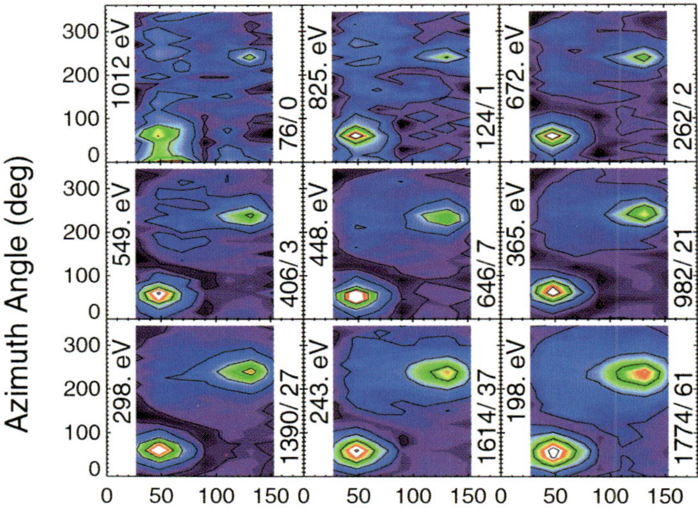

Polar Angle (deg)

Figure 19. Two GEM angle-angle plots showing (a) a typical antisunward solar wind electron strahl and (b) an antisunward as well as a sunward strahl more commonly referred to as a bidirectiional electron stream, often indicative of a coronal mass ejection. The numbers in the left margin of each snapshot give the electron energy sampled while those in the right margin give the maximum and minimum number of counts observed at that energy. The color scale in each frame is normalized to the maximum number of counts seen in that frame.

sample exposure periods. Additionally, the Genesis solar wind data set is quite useful in itself for numerous heliospheric studies. Used in combination with data sets from other spacecraft located L1 (e.g., ACE, WIND, SOHO) and elsewhere (Ulysses) the Genesis solar wind measurements provide outstanding opportunities for multi-spacecraft studies of large-scale phenomena in the solar wind.

Acknowledgements

Work at Los Alamos was performed under the auspices of the U.S. Dept. of Energy with financial support from the NASA Genesis mission. A large number of highly talented individuals at LANL, SwRI, LMA and JPL were involved in making the GEM and GIM instruments a reality and we gratefully acknowledge their invaluable contributions.

References

Bame, S. J., Glore, J. P., McComas, D. J., Moore, K. R., Chavez, J. C., Ellis, T. J., Peterson, G. R., Temple, J. H., and Wymer, F. J.: 1983, 'The ISPM Solar Wind Plasma Experiment', in K. P. Wentzel, R. G. Marsden, and B. Battrick (eds.), *The International Solar Polar Mission – Its Scientific Investigations*, Noordwijk, ESA SP-1050, p. 49.

Bame, S. J., McComas, D. J., Barraclough, B. L., Phillips, J. L., Sofaly, K. J., Chavez, J. C., Goldstein, B. E., and Sakurai, R. K.: 1992, 'The Ulysses Solar Wind Plasma Experiment', *Astron. Astrophys. Suppl. Ser.* **92**, 237.

Burnett, D. S., Barraclough, B. L., Bremmer, R. R., Neugebauer, M., Oldham, L. P., Sasaki, C. N., Sevilla, D., Smith, N., Sansbery, E., Sweetnam, D., and Wiens, R.C.: 2003, 'The Genesis Discovery Mission: Return of Solar Matter to Earth', *Space Sci. Rev.*, this volume.

Gosling, J. T., Asbridge, J. R., Bame, S. J., and Feldman, W. C.: 1978, 'Effects of a Long Entrance Aperture Upon the Azimuthal Response of Spherical Section Electrostatic Analyzers', *Rev. Sci. Instrum.* **49** (9), 1260.

Jurewicz, A. J. G., Burnett, D. S., Wiens, R. C., Friedmann, T. A., Hays, C. C., Hohlfelder, R. J., Nishiizumi, K., Stone, J. A., Woolum, D. S., Becker, R., Butterworth, A.L., Campbell, A. J., Ebihara, M., Franchi, I. A., Heber, V., Hohenberg, C. M., Humayun, M., McKeegan, K. D., McNamara, K., Meshik, A., Pepin, R. O., Schlutter, D., and Wieler, R.: 2003, 'Overview of the Genesis Solar-Wind Collector Materials', *Space Sci. Rev.*, this volume.

McComas, D. J., Bame, S. J., Barker, P., Feldman, W. C., Phillips, J. L., and Riley, P.: 1998, 'Solar Wind Electron Proton Alpha Monitor (SWEPAM) for the Advanced Composition Explorer', *Space Sci. Rev.* **86**, 563.

Neugebauer, M., Steinberg, J. T., Tokar, R. L., Barraclough, B. L., Dors, E. E., Wiens, R. C., Gingerich, D. E., Luckey, E., and Whiteaker, D. B.: 2003, 'Genesis On-Board Determination of the Solar Wind Flow Regime', *Space Sci. Rev.*, this volume.

Nordholt J. E., Wiens, R. C., Abeyta, R. A., Baldonado, J. R., Burnett, D. S., Casey, P., Everett, D. T., Lockhart, W., McComas, D. J., Mietz, D. E., MacNeal, P., Mireles, V., Moses Jr., R. W., Neugebauer, M., Poths, J., Reisenfeld, D. B., Storms, S. A., and Urdiales, C.: 2003, 'The Genesis Solar Wind Concentrator', *Space Sci. Rev.*, this volume.

Wiens R. C., Neugebauer, M., Reisenfeld, D. B., Moses Jr., R. W., and Nordholt, J. E.: 2003, 'Genesis Solar Wind Concentrator: Computer Simulations of Performance Under Solar Wind Conditions', *Space Sci. Rev.*, this volume.

GENESIS ON-BOARD DETERMINATION OF THE SOLAR WIND FLOW REGIME

M. NEUGEBAUER[1], J. T. STEINBERG[2], R. L. TOKAR[2], B. L. BARRACLOUGH[2], E. E. DORS[2], R. C. WIENS[2], D. E. GINGERICH[3], D. LUCKEY[3] and D. B. WHITEAKER[4]

[1]*Jet Propulsion Laboratory, California Institute of Technology, Pasadena, California, U.S.A.*

[2]*Los Alamos National Laboratory, Los Alamos, New Mexico, U.S.A.*

[3]*Lockheed Martin Astronautics, Denver, Colorado, U.S.A.*

[4]*Raytheon Missile Systems, Tucson, Arizona, U.S.A.*

(Author for correspondence, e-mail: mneugeb@cox.net)

Received 25 December 2001; Accepted in final form 14 May 2002

Abstract. Some of the objectives of the Genesis mission require the separate collection of solar wind originating in different types of solar sources. Measurements of the solar wind protons, alpha particles, and electrons are used on-board the spacecraft to determine whether the solar-wind source is most likely a coronal hole, interstream flow, or a coronal mass ejection. A simple fuzzy logic scheme operating on measurements of the proton temperature, the alpha-particle abundance, and the presence of bidirectional streaming of suprathermal electrons was developed for this purpose. Additional requirements on the algorithm include the ability to identify the passage of forward shocks, reasonable levels of hysteresis and persistence, and the ability to modify the algorithm by changes in stored constants rather than changes in the software. After a few minor adjustments, the algorithm performed well during the initial portion of the mission.

1. Introduction

The purpose of the Genesis mission is to determine the elemental and isotopic compositions of the outer layers of the Sun. The method used is to collect samples of the solar wind and return those samples to Earth for detailed analyses in state-of-the-art laboratories (Burnett *et al.*, 2003). But although the solar wind originates in the outer layers of the Sun, some change in composition (fractionation) occurs during the processes of removing that material from the Sun and accelerating it into the solar wind. The strength and nature of the fractionation are not constant, but depend on the variable properties of the solar wind. These variations are best characterized by considering three different types, or regimes, of solar-wind flow. There are two types of quasi-stationary solar wind – the high-velocity streams from the cooler, less dense regions of the solar atmosphere called coronal holes (CH) and the slower interstream (IS) flow observed between successive CH streams. The quasi-stationary CH and IS flows are occasionally interrupted by the transient injection of new material into the solar wind in explosive events called coronal

Space Science Reviews **105:** 661–679, 2003.

mass ejections (CME). CMEs occur more frequently at high activity phases of the solar cycle than at solar minimum.

Elemental fractionation of the solar wind depends largely on the time required for an element to become ionized in the solar atmosphere. The ionization time, in turn, is most strongly influenced by an element's first-ionization potential (FIP = the energy required to remove one electron from an atom). The elemental fractionation due to the FIP effect is observed to be much greater in IS than in CH flow; in IS wind, low-FIP elements are enhanced compared to high-FIP elements by a factor of 4 to 5, whereas the enhancement factor is $\leqslant 2$ for CH wind (von Steiger *et al.*, 2000). A second, much weaker source of fractionation is Coulomb drag close to the Sun where the solar wind still undergoes collisions; this source of fractionation depends on the mass/charge ratio of the ions. Because each isotope of an element has the same FIP, the isotopic fractionation is expected to be much less than the elemental fractionation, and may even be negligible in the quasi-stationary solar wind. The fractionation properties of the CME wind are much more variable than those of the CH and IS winds. Large enhancements of the ^{3}He:^{4}He isotope ratio have been observed in several CME events (Gloeckler *et al.*, 1999; Ho *et al.*, 2000). Elemental abundance ratios are also highly variable in CMEs (Galvin, 1997).

To be able to correct for elemental fractionation and to test for any isotopic fractionations in elements heavier than He, Genesis will expose different collectors to CH, IS, and CME solar wind. The challenge is to determine the correct regime in real-time from the information provided by the Genesis solar wind monitors (described in the paper by Barraclough *et al.*, 2003). This paper describes the on-board algorithm developed to make such a determination. Section 2 explains the logic of the algorithm. Section 3 describes the method of calculating the parameters needed as input to the algorithm, while Section 4 provides some insight into the performance of the algorithm during the first few months after launch.

2. The Algorithm

The data available as input to the regime-selection algorithm are obtained by the Genesis Ion Monitor (GIM) and the Genesis Electron Monitor (GEM). The design and operation of the two monitors are described in the companion paper by Barraclough *et al.* (2003). Ideally, the identification of solar-wind regimes would be based not only on the parameters that can be measured by GIM and GEM, but also on data from other types of instruments, such as a magnetometer, an ion mass spectrometer, and energetic particle detectors. But there are several identifying characteristics of each regime that are within the capability of the instrumentation on Genesis. Coronal hole (CH) wind is characteristically fast (speed > 500 km s^{-1}) and hot (proton temperature $> 1-2 \times 10^5$ K); it has a generally steady helium abundance of 4 to 5%; and suprathermal electrons stream outward from the Sun along the interplanetary magnetic field. Interstream (IS) wind is slower (speed usually

< 450 km s^{-1}) and cooler (proton temperature $\leq 10^5$ K) than the CH wind; its helium abundance is generally lower (0–4%) and more variable; and suprathermal electrons usually flow away from the Sun, but sometimes they are absent. The coronal mass ejection (CME) wind can have any speed; its proton temperature is often anomalously low due to the 3-dimensional expansion as the ejecta flow away from the Sun; its helium abundance is sometimes strongly enhanced (8–30%); and suprathermal electrons often stream in both directions along the magnetic field, which is believed to indicate that the field has the form of a large loop anchored in the Sun.

Reflecting these nominal properties of the different flow types, the regime algorithm is based on estimates of the following plasma parameters calculated following each 2.5-min GIM/GEM data cycle:

$V_p =$ proton speed,

$N_p =$ proton number density,

$T_p =$ proton temperature,

$N_a =$ alpha-particle number density,

$B_e =$ measure of bidirectional electron streaming,

The detailed definitions and the method of calculating these parameters are described in Section 3.

In addition to the capabilities and constraints provided by GIM and GEM, there are several additional requirements on the regime algorithm. First the number of regime changes must be minimized consistent with the goal of collecting sufficiently pure samples of each regime. The motors that run the array-changing mechanisms have a design requirement of 400 regime changes over the course of the mission, although the motors are expected to last much longer than that. Furthermore, each array change requires up to 6 min to implement and can disturb the pointing of the spacecraft for up to an hour. This low-change-rate requirement means that the algorithm should have both hysteresis and persistence. A second requirement is that the algorithm should be biased toward minimizing the contamination of the CH sample by CME or IS wind; the CH sample is given special protection because its composition is expected to be less fractionated than the IS and CME winds. A third requirement is that the algorithm be adjustable largely by changing numerical parameters in look-up tables as opposed to uploading new software.

The first step in determining the correct regime is to form running averages of the parameters V_p, T_{ex}/T_p, N_a/N_p, and B_e. The nominal averaging duration t_{avg} is one hour, but can be lengthened up to two hours by ground command. The parameter T_{ex} is the proton temperature expected for quasi-stationary (CH and IS) wind with speed V_p. The ratio T_{ex}/T_p should average about 1 for CH and IS wind, but significantly exceed unity for CME wind. The calculation of T_{ex} is described in Sections 3 and 4. Running averages of V_p, N_p, and T_p are also calculated over the first and second halves of the interval t_{avg}. V_1 and V_2 are the first and second $t_{avg}/2$ values of the speed, and N_1, N_2, T_1, and T_2 are similarly defined averages of proton density and temperature.

In addition to addressing the persistence requirement, the averaging allows continued operation of the logic even when some parameters are temporarily missing. All ion data (V_p, T_p, N_p, and N_a) will be missing once every 20 data cycles when GIM performs its routine search mode, or when abrupt changes in speed trigger an additional search mode (Barraclough *et al.*, 2003). When T_p is unusually high, the alpha-particle peak in the spectrum cannot be discerned above the proton distribution, so N_a cannot be calculated. The GEM bidirectional streaming parameter B_e can be disabled when the angular distribution of electrons indicates that the spacecraft may be magnetically connected to the Earth's bow shock or when the GEM background becomes high during solar energetic particle events. The drawback of the averaging process is that whenever the regime changes, the wrong collector will be exposed for a time $t_{avg}/2$ prior to an array change. For 150 regime changes/year and t_{avg} = one hour, the wrong collector would be exposed an average of 0.9% of the time even if the algorithm otherwise worked perfectly.

The algorithm for determining the regime depends on whether or not a forward shock has recently passed the spacecraft. Shock passage is deemed to have occurred when the following conditions are met:

$$V_2 - V_1 > V_{jump}, \tag{1}$$

$$N_2/N_1 > R_N, \tag{2}$$

$$T_2/T_1 > R_T. \tag{3}$$

Nominal values of the constants are $V_{jump} = 40$ km s^{-1}, $R_N = 1.4$, and $R_T = 1.5$. These values were chosen to identify strong shocks with a very low rate of false alarms; weak shocks will not be detected. The time of the shock passage ($t_{shock} = t_{now} - t_{avg}/2$, where t_{now} is the current time) is saved.

The logical flow of the regime selection algorithm is diagrammed in Figure 1. At the start of the analysis, the current regime is denoted by a parameter LASTYPE = 0 (instrument startup), 1 (CME), 2 (CH), or 3 (IS). Suppose that LASTYPE $\neq 1$; i.e., the regime is not CME. Whether or not the regime should be changed to CME depends on the parameter TOCME, whose derivation is shown in Figure 2. The CT*n*, CA*n*, and CB*n* are adjustable constants. Because, at a solar distance of 1 AU, almost all interplanetary shocks are driven by CMEs, we make it easier to enter the CME regime within a fixed time interval after t_{shock}. The boundaries of that time interval are determined by the uploadable parameters *ctime1* and *ctime2*, whose nominal values are 5 and 25 hours, respectively. Because alpha-particle and bidirectional electron streaming data will sometimes be missing, those parameters are assigned weights, W_a and W_b, respectively, which are 1 if there are data within

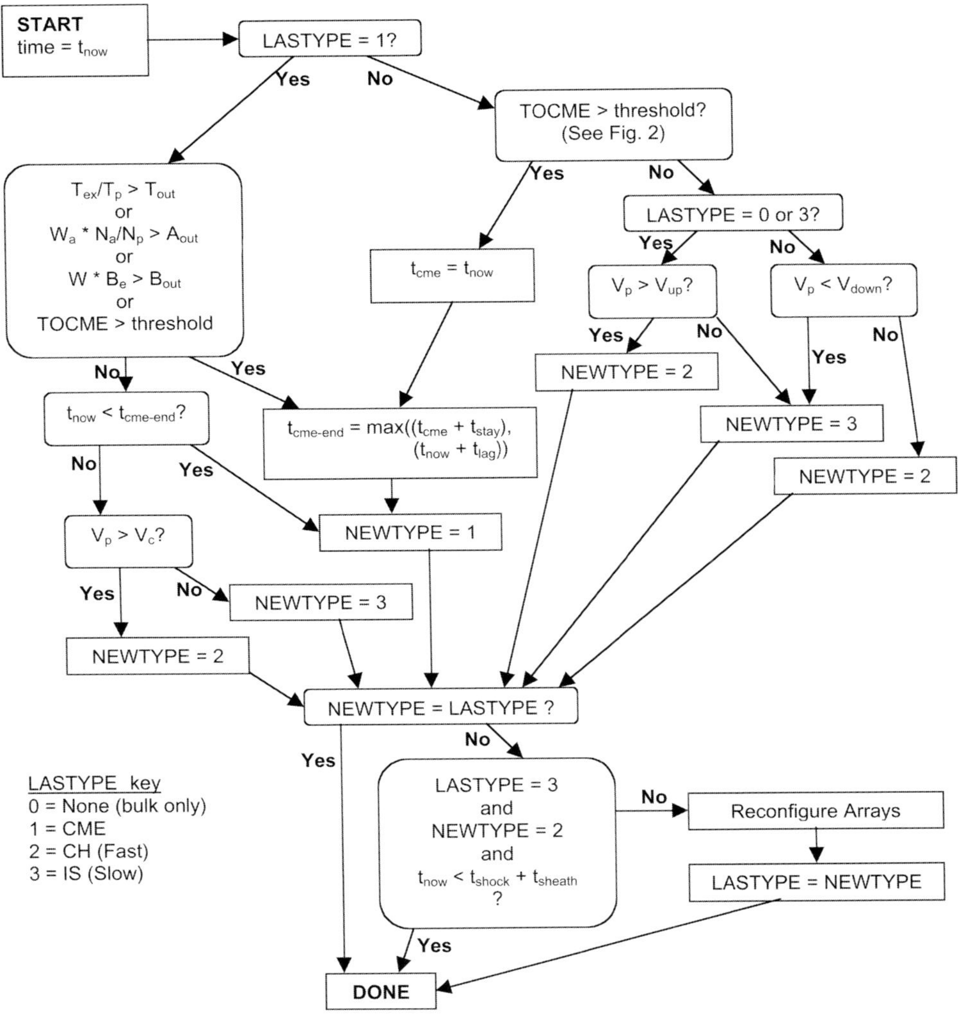

Figure 1. Logical flow of the regime-selection algorithm

the interval t_{avg}, and 0 if there are not. (No analysis is done if there are no values of either V_p or T_p within the time t_{avg}.) A simple fuzzy-logic scheme is used to combine the parameters T_{ex}/T_p, N_a/N_p, and B_e. As an example, Figure 3 shows how, for CA1 = 23, CA2 = 1.15, CA3 = 16.67, and CA4 = 1.0, the alpha-particle parameter Y_a depends on N_a/N_p for intervals that do and do not follow a shock. Similar diagrams could be drawn for Y_t and Y_b. TOCME is calculated from the weighted averages of Y_a, Y_t, and Y_b. If TOCME exceeds a threshold value (currently set at 0.4), the time of entering the CME is noted (as t_{cme}), LASTYPE is set to 1, and the collector arrays are changed to expose the CME array.

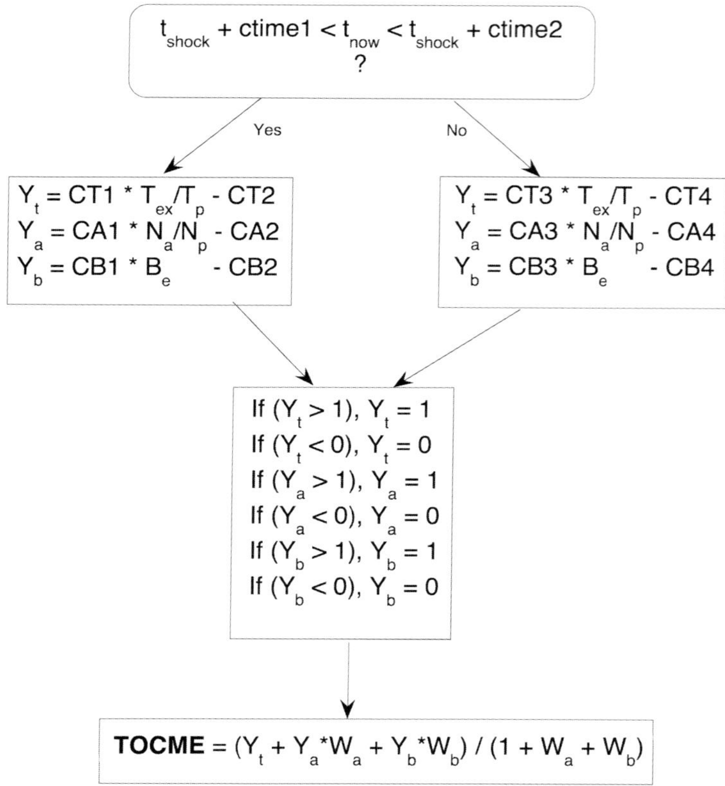

Figure 2. Expanded view of the logic that combines the proton temperature, helium abundance, and bidirectional streaming parameters to determine the likelihood of wind from a coronal mass ejection. The W_a and W_b are relative weights of the helium abundance and bidirectional streaming parameters based on the number of valid determinations over the averaging interval.

It is especially important not to allow the parameter Y_b alone to trigger a change into the CME regime. This is because true bidirectional electron streaming is sometimes found upstream of both forward and reverse shocks in corotating interaction regions (Gosling *et al.*, 1993) and because depletions in suprathermal electrons are sometimes observed at pitch angles of 90° (Gosling *et al.*, 2001) and these depletions can be mistaken for CME-associated bidirectional streaming. We guard against false entry into the CME regime in two ways: (1) For the TOCME threshold greater than $\frac{1}{3}$, more than one of the Y parameters must exceed 0 to trigger entry to the CME regime; e.g., if $Y_t = 1$ with Y_a and $Y_b = 0$, TOCME $= 0.33$ and the regime is not changed (2) Use of the sum $(2 + W_b)$ in the denominator of the expression for TOCME in Figure 2 (rather than $1 + W_a + W_b$) prevents bidirectional electron streaming alone from causing entry into CME mode even when the alpha-particle density could not be determined and $W_a = 0$.

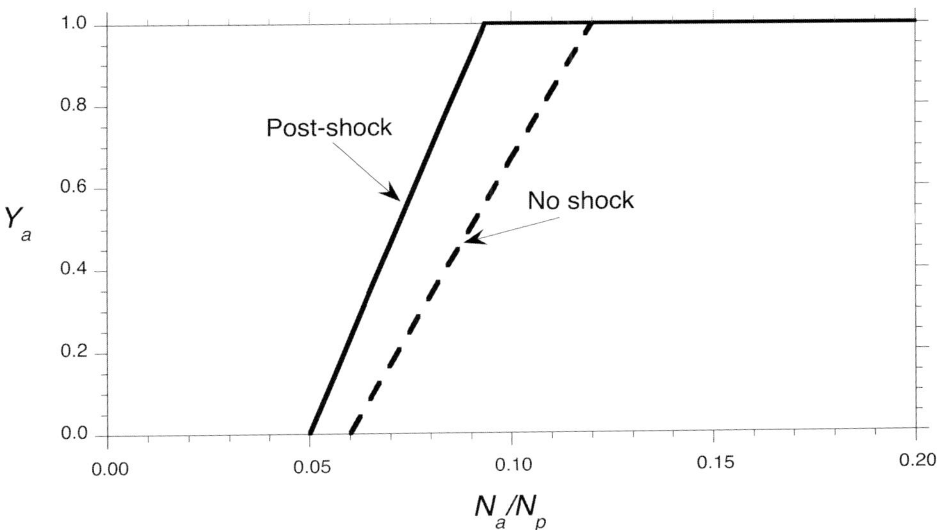

Figure 3. Illustration of the calculation of the parameter Y_a (used in Figure 2) from the helium abundance ratio N_a/N_p.

The trailing edges of CMEs are less well defined than are their leading edges (e.g., Neugebauer *et al.*, 1997). To protect the CH and IS collectors from CME wind, we make it harder to exit from the CME regime than it was to enter. The CME signatures often come and go within an event. The high helium abundance signature is notoriously patchy (Zwickl, 1983) and the bidirectional electron streaming feature can disappear as some of the solar field lines reconnect with the interplanetary magnetic field (Gosling *et al.*, 1995). For these reasons we require that a minimum time, t_{stay}, be spent in the CME regime. The current value of t_{stay} is 18 hours. After that time has elapsed, the CME regime is retained as long as any of the three signatures (N_a/N_p, T_{ex}/T_p or B_e) remains above their respective threshold values; currently, $A_{\text{out}} = 0.06$, $T_{\text{out}} = 1.5$, and $B_{\text{out}} = 0.4$. Persistence is provided by requiring the non-CME conditions to prevail for a time t_{lag} (currently 6 hours) before switching to either CH or IS.

Upon exiting the CME regime, the new regime is determined on the basis of the proton speed. If $V_p > V_c$, the new regime is CH; otherwise it is IS. Currently, V_c is set at 500 km s^{-1}.

Next, suppose Genesis is in the CH regime (LASTYPE = 2), TOCME is less than threshold, and the velocity starts decreasing. When V_p drops below V_{down} (currently 425 km s^{-1}), the regime changes to IS. It cannot return to the CH regime unless V_p subsequently increases above V_{up} (currently 525 km s^{-1}). This gap between V_{up} and V_{down} provides the required hysteresis. There is, however, a second requirement on changing from IS to CH to account for the fact that slow, IS wind that has been accelerated by an interplanetary shock has its source in the interstream wind, not in coronal holes. When an interplanetary shock is detected in an IS regime, the IS regime is held until either a CME is encountered or a time t_{sheath} (currently 12 hours) has elapsed.

3. Calculation of parameters used by the algorithm

As explained in the accompanying paper by Barraclough *et al.* (2003), GIM acquires a 3-dimensional array of counts at 40 energy channels (voltage levels) in 40 azimuth directions (angle φ) for each of 8 detectors (angle θ) every 2.5 min. These 12,800 count values are then checked for reasonableness and corrected for dead time and background level; the detailed logic of those operations is beyond the scope of this review. A three-dimensional phase space density $f_{3D}(v_p,\varphi,\theta)$ is then calculated in the usual way as $f_{3D}(v_p,\varphi,\theta) = 2\times\text{counts}/(v_p^4 G\Delta\text{t})$, where $v_p = \sqrt{2(E/q)/m_p}$, m_p is the proton mass, E/q is the proton energy per unit charge, which is proportional to the voltage applied to the analyzer, G is a geometric factor, and Δt is the time over which counts are accumulated. At this stage of the calculation it is assumed that all counts are due to protons. Next, the matrix of 12,800 3-D values of phase space density are summed over all angles to yield a one-dimensional distribution function

$$f_{1\text{D}}(v_p) = \sum\sum v_p^2 f_{3\text{D}}(v_p, \varphi, \theta)\ \sin(\theta)\Delta\varphi\Delta\theta. \tag{4}$$

This 1-D spectrum is then divided into a proton spectrum extending from $v_p = v_L$ to v_M and an alpha-particle spectrum extending from $v_p = v_M$ to v_H. Without going into the details, this process includes steps to ensure that the velocity bin with the highest value of $f_{1\text{D}}(v_p)$ is not an outlier, that a second (alpha-particle) peak exists at a reasonable number of velocity bins above the highest (proton) peak, and that the height of the alpha-particle peak is sufficiently above the minimum value between the two peaks. The limits v_L and v_H are the lowest and highest proton speeds in the 1-D spectrum and v_M is the proton speed at the minimum between the proton and alpha peaks. When it is not possible to find a reliable alpha-particle peak, v_M is set to a value of v_p 27% higher than the value of v_p for which $f_{1\text{D}}(v_p)$ reaches its peak, or that value for which $f_{1\text{D}}(v_p)$ is down to 2.5% of its peak, whichever is smaller.

Moments are then calculated from the 1-D distribution function as follows:

$$N_p = \int_{v_L}^{v_M} f_{1\text{D}}(v_p)\,\text{d}v_p, \tag{5}$$

$$N_a = \sqrt{2}\int_{v_M}^{v_H} f_{1\text{D}}(v_p)\,\text{d}v_p, \tag{6}$$

$$V_p = [N_p]^{-1} \int_{v_L}^{v_M} v_p f_{1D}(v_p)\, dv_p, \tag{7}$$

$$T_p = [m_p/kN_p] \int_{v_L}^{v_M} [v_p - V_p]^2 f_{1D}(v_p)\, dv_p. \tag{8}$$

The resulting values of N_p, N_a, V_p, and T_p are then checked against uploadable sanity limits; for example, V_p must be between 200 and 1200 km s^{-1} for the data for that cycle to be included in the averages. As a final step in quality control, the moments' histories are despiked by the removal from the running averages of data from a single data cycle whose values are inconsistent with recent parameter trends.

In the quasi-stationary wind, the proton temperature increases with the proton speed. That unusually low temperatures are associated with transient solar wind events was recognized by Gosling *et al.* (1973). Richardson and Cane (1995) demonstrated that the decrease of the temperature well below a mean expected value (called T_{ex}) is a useful quantitative indicator of CMEs. A number of different fits have been made to different data sets to derive T_{ex} as a function of V_p. T_{ex} has variously been fit to a quadratic in V_p (Burlaga and Ogilvie, 1973), a linear function of V_p (Neugebauer *et al.*, 1997), or quadratic at low speed and linear at high speed ($>$ 500 km s^{-1}) (Lopez, 1987). To retain flexibility, we assume a relation of the form, for $V_p < V_{\text{cut}}$:

$$T_{ex} = C_1 + C_2 V_p + C_3 V_p^2 \tag{9}$$

and, for $V_p > V_{\text{cut}}$:

$$T_{ex} = C_4 + C_5 V_p + C_6 V_p^2 \tag{10}$$

At the start of the mission, we followed Richardson and Cane and used the constants derived from the relation developed from the NASA OMNI data set by Lopez (1987), for which $V_{\text{cut}} = 500$ km s^{-1}, $C_1 = 26,000$, $C_2 = 316.2$, $C_3 = 0.961$, $C_4 = -142,000$, $C_5 = 510$, and $C_6 = 0$ for V_p in km s^{-1} and T_{ex} in units of 10^3 K. The form of this relationship is shown as the curve marked RC in Figure 4. As explained in Section 4, the constants were modified after analysis of the Genesis flight data.

The bidirectional electron streaming parameter B_e is calculated from the data acquired by GEM (Barraclough *et al.*, 2003). During a 2.5-min data cycle, GEM obtains counts in each of 7 CEMs (polar angle) in 40 azimuth bins and 24 energy channels. The search for bidirectional streaming is limited to only an intermediate energy range for several reasons. The character of the core electron population at lower energies may be due to local effects rather than to the global field topology. At the highest energies the electrons provide too few counts per data cycle to yield

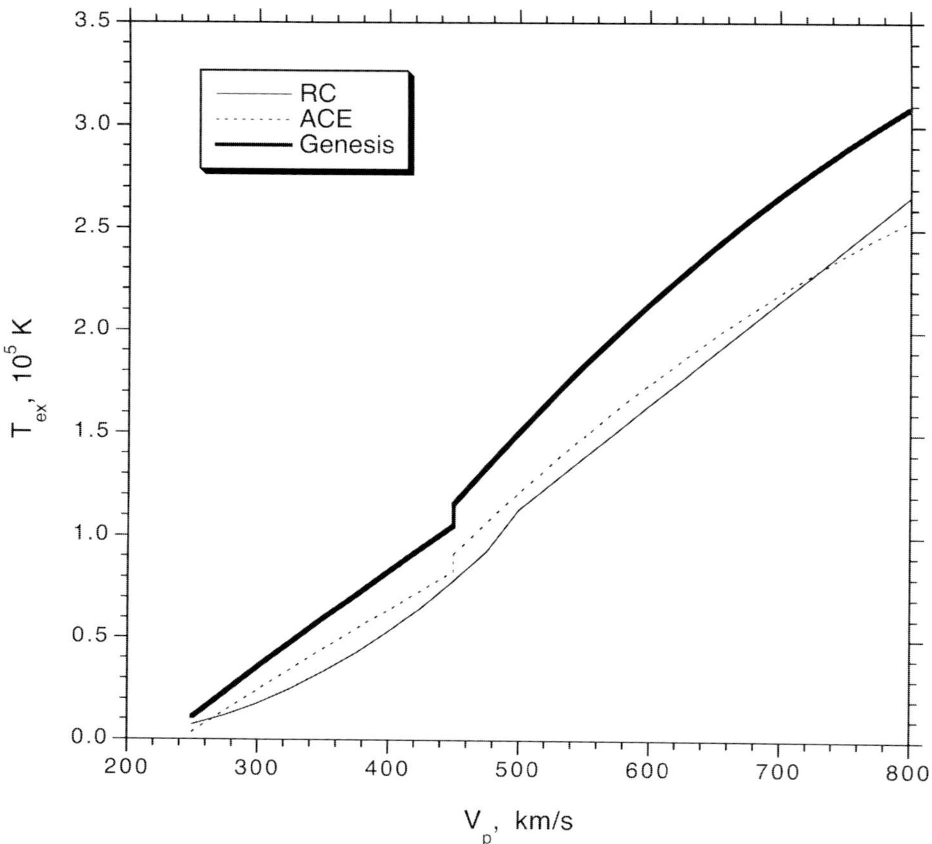

Figure 4. Comparison of three expressions for the dependence of expected proton temperature T_{ex} on solar wind speed. RC is the expression used by Richardson and Cane for identification of CMEs. ACE is the expression derived from three-years of ACE data. The heavy curve is the value used for Genesis based on the ACE T_{ex}-V curve adjusted for the fact that Genesis measures higher proton temperatures than does ACE.

statistically significant results. The energy range to be used can be changed by ground command, but a nominal range is the 12 energy channels between 160 and 984 eV. The counts are first normalized by the geometric factor for each CEM and then summed over all CEMs to yield a 2-dimensional matrix in energy and azimuthal angle. Then, for each of the energy levels in the range of interest, the computer finds (1) the azimuth bin A_{peak} with the greatest number of counts C_{peak}, (2) C_{min}, which is the smaller of the number of counts averaged over bins near $+90°$ and $-90°$ from A_{peak}, and (3) C_{180}, which is the number of counts 180° from A_{peak}. Bidirectional streaming is deemed to be present at a given energy level if both $C_{\text{peak}}/C_{\text{min}}$ and C_{180}/C_{min} exceed uploadable threshold values. A positive detection of bidirectional electron streaming (BDE) requires the presence of BDE behavior as outlined above in at least a minimum number of energy levels (nominally 3 or 4).

If the interplanetary magnetic field is approximately aligned with the spacecraft spin axis, bidirectional streaming would be evident in the polar, rather than in the azimuthal angular distribution. Therefore, if no BDE is found in the analysis of the azimuthal angles, a second search is done using the polar angles. The counts are again normalized according to geometric factor and then summed over azimuth for each relevant energy level. $C_{\min}$ is found by averaging over an uploadable list of CEMs (typically CEM#3 to CEM#5) and bidirectional streaming is deemed to be present for that energy level if both (Counts in CEM#1)/$C_{\min}$ and (Counts in CEM#7)/$C_{\min}$ exceed uploadable thresholds. As with the azimuthal search, BDEs must be detected in more than a minimum number of energy levels.

If bidirectional streaming is detected by analysis of either the azimuthal or the polar data, the parameter $B_e = 1$; otherwise, $B_e = 0$. When the interplanetary magnetic field connects the spacecraft to the region of the Earth's bow shock, bidirectional streaming may be caused by backstreaming of electrons accelerated at the shock, and thus give a false indication of CME conditions. For this reason, the BDE analysis can be disabled (the B_e weight W_b can be set to 0) if the polar direction of Earth is less than a threshold value (nominally 22°).

About twice a week during routine operations, the calculated moments and B_e together with compressed values of all the electron counts and a masked set of the ion counts for each data cycle are telemetered to the ground. Ground-based analyses determine the desirability of changing any of the parameters used in the on-board analyses. It is possible to adjust the computation of the moments and B_e to avoid using the counts acquired by any CEM that malfunctions.

4. In-Flight Performance

The regime-selection algorithm described in the previous sections was developed, and to the extent possible tested, using data from the ISEE-3 solar wind spectrometer measurements in 1978–1980. Additional, but necessarily limited testing was carried out before launch using dummy data. The first real test occurred during the first few months of the Genesis mission before the solar-wind collection began.

Following Genesis launch on August 8, 2001 and the spacecraft checkout period, GEM was turned on August 23 and GIM was turned on August 24. Since then, the operation of the monitors has met all expectations. During several large solar proton events, the background counts in GEM were too high (>1000 counts per bin) to allow calculation of B_e, but the GIM background was much lower and the ion data were still usable.

The regime algorithm was activated on August 25. An error in the on-board geometric factors for GEM which prevented the proper operation of the search for bidirectional electron streaming was discovered and then corrected on September

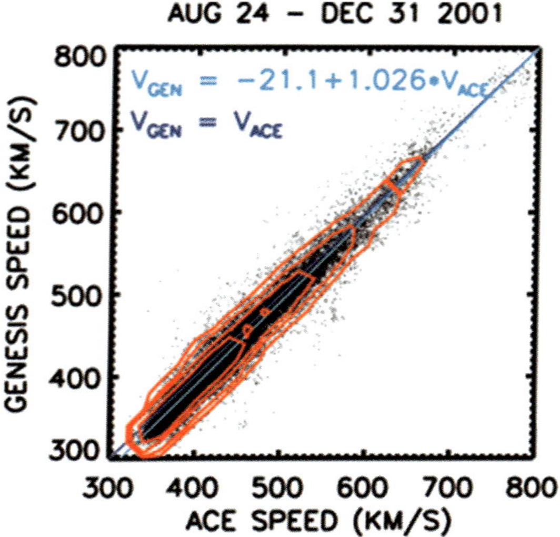

Figure 5. Scatter plot of the nearly simultaneous proton speeds observed by Genesis and ACE.

19. On October 17, adjustments were made to the part of the algorithm that separates the alphas from the protons in the 1-D spectrum to improve the rejection of alpha-particle densities when the protons are too hot for a reliable identification of the alpha peak.

During the check-out period some excessive switching in and out of the CME regime was caused by the ephemeral nature of the CME signatures. This led to the addition of the parameter t_{lag} to the logical sequence as shown in Figure 1. This software change was implemented on December 13.

The values of the plasma moments calculated on-board from the 1-dimensional distribution functions have been compared to the moments calculated in the conventional manner from 3-dimensional data acquired by the SWEPAM instrument on the ACE spacecraft (McComas *et al.*, 1998). Figure 5 is a scatter plot of the proton speed calculated on Genesis versus the proton speed determined by ACE for the period August 24 to December 31, 2001. Only intervals with V_p(ACE) $\geqslant 350$ km s^{-1} were included, thereby avoiding intervals for which ACE measurements have a somewhat larger uncertainty; intervals affected by solar energetic particles were also excluded. The time resolutions are 2.5 min for Genesis and 64 s for ACE. ACE measurements were interpolated to Genesis times, but no correction was made for the time shift (varying from 0 to 60 min) between features propagating past the two spacecraft. The correlation is excellent ($R = 0.981$), and the on-board calculation of the proton speed is deemed to be entirely adequate to meet the purposes of the regime algorithm.

Figure 6 shows a similar plot for the proton temperatures measured by Genesis and ACE. The Genesis temperatures are systematically higher than the ACE temperatures. The Genesis temperatures also appear to have a lower limit close

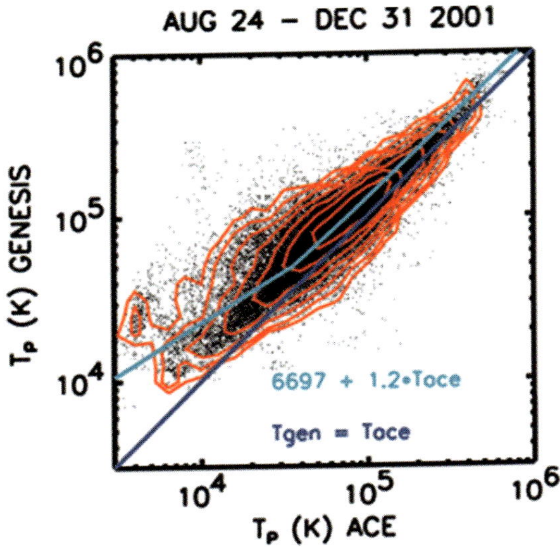

Figure 6. Scatter plot of the nearly simultaneous proton temperatures observed by Genesis and ACE.

to 10^4 K. The difference is caused by the different methods of calculation. On-board Genesis, the summations in Equation (4) extend over all angles, even those far from the part of the field of view that contains the peak of the distribution. Calculation of the ACE temperature, however, is limited to the part of phase space surrounding the peak of the distribution and is therefore less affected by the tail of the distribution and background counts. A result of the high temperatures calculated from Equations (4) and (8) is that T_{ex}/T_p was seldom greater than unity, even in some periods that were obviously CMEs. This situation was corrected by generating a new set of constants in Equations (9) and (10). First we fit 3 years of ACE speed and temperature data to find a T_{ex} relation for ACE; the results are given by the dashed curve in Figure 4. The data in Figure 6 were then fit to the relation $T_{\mathrm{Genesis}} = 1.214 \times T_{ACE} + 6697$, and that relation was used to modify the T_{ex} calculated for ACE to yield a $T_{ex} - V_p$ relation for Genesis. The resulting curve is shown as the heavy line in Figure 4. The new constants in Equation (9) and (10) are $V_{\mathrm{cut}} = 450$ km s^{-1}, $C_1 = -127\,800$, $C_2 = 595.2$, $C_3 = -0.1623$, $C_4 = -324\,400$, $C_5 = 1217$, and $C_6 = -0.5214$ for V_p in km s^{-1} and T_{ex} in K.

The Sun was remarkably active in late 2001, resulting in many CMEs and interplanetary shocks. The Genesis shock-detection algorithm given in Equations (1) to (3) worked very well without any adjustments. Table 1 compares the shocks detected on-board Genesis with the shocks detected by the Proton Monitor of the Celias experiment on the SOHO spacecraft. (The SOHO shock list is available at the University of Maryland website at http://umtof.umd.edu/pm/). The Genesis shock passage times given in this table were obtained from the high-resolution (2.5-min) data obtained during the time t_{avg} preceding the on-board detection of the shock. Fourteen shocks were detected by both spacecraft. Genesis missed one weak

TABLE I
Times of interplanetary shocks.

Date	UT time	
	SOHO	Genesis
08/27/01	19:14	19:30
08/30/01	13:26	13:38
09/01/01	Not listed	02:33*
09/14/01	01:08	01:22
09/25/01	19:51	20:09
09/29/01	09:03	09:10
09/30/01	18:41	18:48
10/11/01	16:13	16:25
10/21/01	16:05	16:13
10/25/01	07:57	08:29
10/28/01	02:33	02:38
10/31/01	12:49	13:14
11/06/01	01:20	Monitors Off
11/06/01	Not Listed	16:30[2]
11/15/01	13:45	Not Detected[3]
11/19/01	17:34	17:46
11/24/01	05:33	Monitors Off
12/17/01	Not Listed	03:07[4]
12/29/01	04:56	04:58
12/30/01	19:38	19:39

[1]Looks like a weak shock in the plasma parameters, but the ACE magnetometer shows the field strength decreasing rather than increasing. Probably a bad detection.
[2]Not a clean shock signature; embedded within a fast CME.
[3]Weak shock not found by shock-detection algorithm but visible in data.
[4]Appears to be a forward wave which has not steepened into a shock.

shock that was observed by SOHO. No SOHO shocks corresponded to three of the Genesis shocks; at least two of those three were probably erroneous detections. Through November, the times of the Genesis shock passages followed the SOHO passages by 5 to 32 min because Genesis was still moving out to the L1 point and was downstream of SOHO. For the December shocks, both spacecraft were in halo orbits around L1 and the times agreed within the 0.5-min resolution of SOHO and the 2.5-min resolution of Genesis. Further studies of shocks and other structures in the solar wind will be carried out through more detailed comparisons of data from Genesis, SOHO, and ACE.

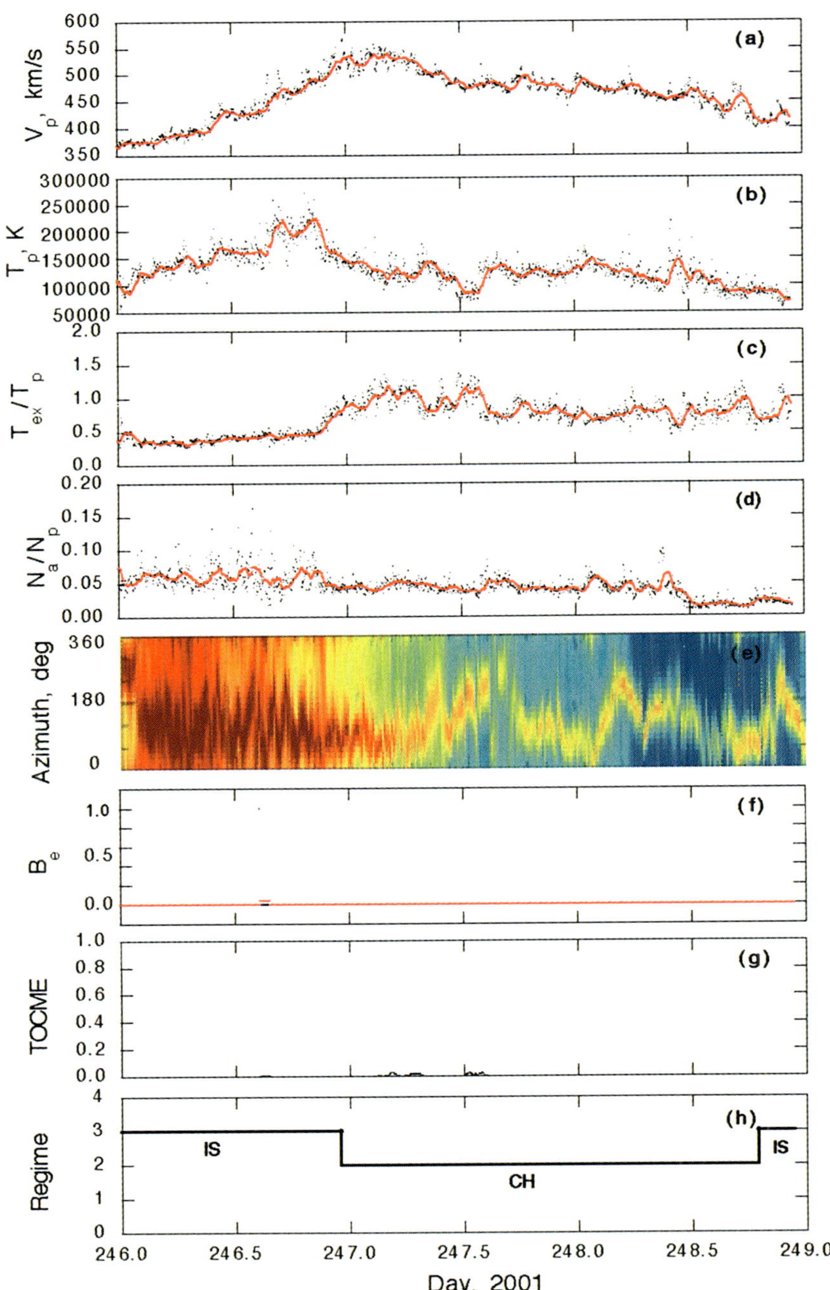

Figure 7. Example of Genesis identification of a solar wind stream from a coronal hole. From top to bottom are plotted (a) proton speed V_p, (b) proton temperature T_p, (c) the ratio of the expected temperature to the observed temperature, (d) the helium abundance ratio N_a/N_p, (e) the azimuthal angular distribution of electrons in the energy range 243 to 298 eV, (f) the bidirectional streaming parameter B_e, (g) the TOCME parameter, and (h) the regime selected. In each panel except (e), the black points are the instantaneous 2.5-minute values and the red points are the hourly running averages.

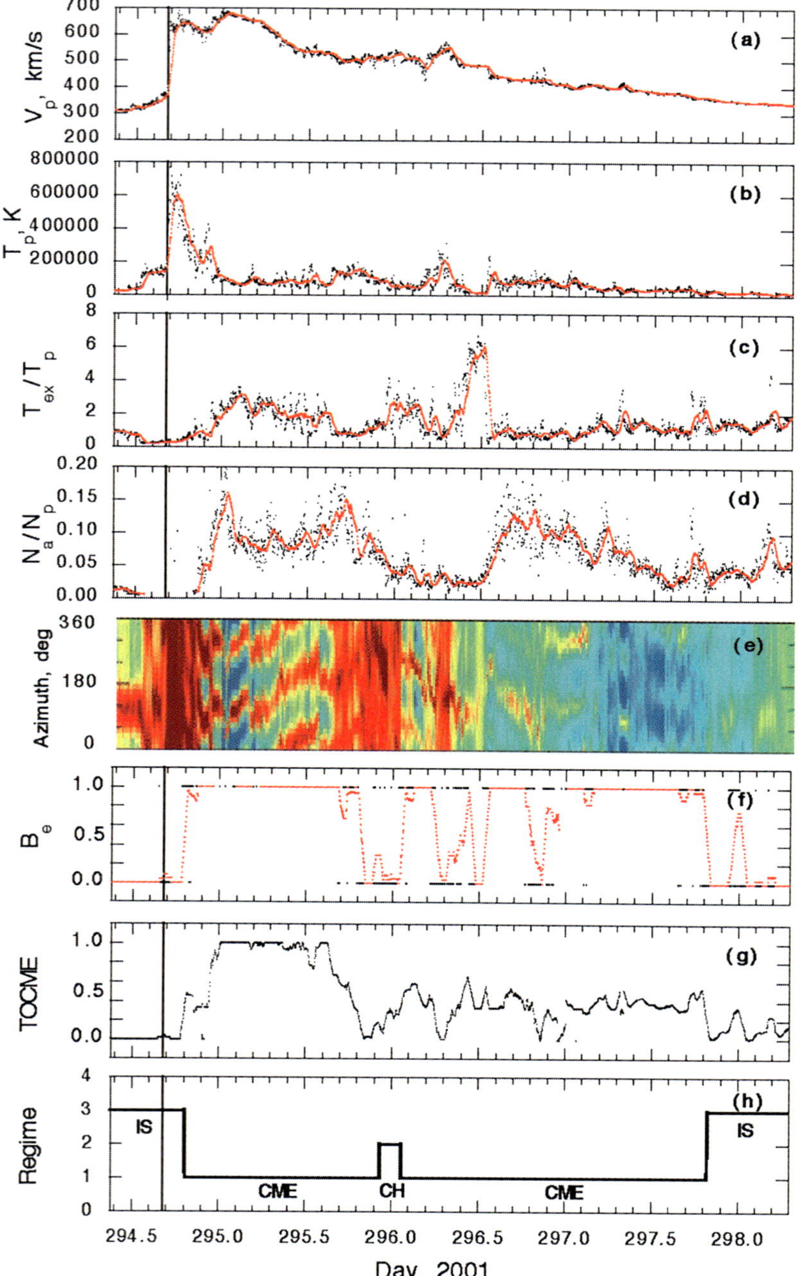

Figure 8. Example of Genesis identification of a solar wind stream associated with a coronal mass ejection. The panels are the same as those in Figure 7.

Figures 7 and 8 provide examples of the operation of the regime algorithm during two different intervals. From top to bottom, the panels show (a) V_p, (b) T_p, (c) T_{ex}/T_p, (d) N_a/N_p, (e) the azimuthal distribution of suprathermal electrons over the energy range 243 to 298 eV, (f)vB_e, (g) the parameter TOCME, and (h) the regime deduced from the data. Individual 2.5-min values are plotted in black and the running hourly averages are plotted in red. The data shown in panel (c) are the values of T_{ex}/T_p based on the Richardson-Cane equation for T_{ex} in use at the time the data were acquired; the revised equation for T_{ex}, which was not yet in use, would be higher.

Figure 7 shows the entry into and exit from a rather small CH stream. No shock was detected during that interval. For most of the CH interval, the helium abundance ratio N_a/N_p agreed with expectations for coronal hole flow, thereby corroborating our choice of a value for V_{up}. During the early Genesis period, there were no CH streams with the long durations and very high speeds typical of periods of declining solar activity.

A period that includes a CME is shown in Figure 8. Detection of a forward shock on day 294 is indicated by a vertical line. The regime before the shock was IS and it remained IS despite the high speed because of the $t_{\mathrm{sheath}} = 12$ hr requirement. The regime changed to CME near day 294.8 on the basis of a strong onset of bidirectional electron streaming together with a small rise in T_{ex}/T_p. A helium enhancement and a much stronger temperature depression (high T_{ex}/T_p) soon followed. At day 295.93, the regime changed from CME to CH on the basis of the running averages of $T_{ex}/T_p = 1.42$, $N_a/N_p = 0.069$, $B_e = 0.17$, and $V_p = 516$ km s^{-1}; the values of T_{out}, A_{out}, and B_{out} in use at the time were 1.5, 0.07, and 0.4, respectively. T_{ex}/T_p started to rise almost immediately thereafter, but TOCME did not reach the threshold value of 0.4 required to reenter the CME regime until day 296.06. In retrospect, we think that CME conditions may have prevailed essentially continuously from day $\sim$294.8 until day $\sim$297.8. The algorithm would have held the CME regime without the short excursion into CH if the t_{lag} part of the logic had been in place at the time.

5. Conclusions

We have developed and flight-tested an algorithm for determining the flow regime (coronal hole, interstream, or coronal mass ejection) of the solar wind on the basis of running hourly averages of solar wind ion and electron parameters. After a few adjustments, the performance of the algorithm has turned out to be generally satisfactory While the algorithm now makes few blatant errors, its success rate is difficult to quantify at this point.

To protect the cleanliness of the CH bin we have modified some parts of the software and some parameters to tighten up the exclusion of CME wind from the CH regime. We will continue to monitor the regime idenfications throughout the mission, not only by examining the Genesis GIM and GEM data alone, but by comparing it to nearly simultaneous data acquired by ACE. The ACE composition data are expected to be particularly useful in identifying stream interfaces between IS and CH flows. As a result of further analyses, we may adjust the parameters V_{up} or V_{down} to minimize mixing of the CH and IS regimes. At the end of the mission we will estimate the cross-contamination of each of the regimes by the other two. Because CMEs are often difficult to identify even under the best of circumstances with many more data types than are available on Genesis, perfection should not be expected.

We note in closing that similar algorithms might be useful in several other contexts.

Acknowledgements

Part of this work was performed at the Jet Propulsion Laboratory (JPL) of the California Institute of Technology under a contract with the National Aeronautics and Space Administration (NASA). The work at Los Alamos was performed under the auspices of the U. S. Department of Energy under NASA contract W-19,272. The work at Lockheed Martin was performed under a contract from JPL. We thank F. Ipavich for providing the Celias proton monitor data from the Solar Heliospheric Observatory (SOHO); SOHO is a joint European Space Agency and NASA mission.

References

Barraclough, B. L. *et al.*: 2003, 'The Plasma Ion and Electron Instruments for the Genesis Mission', *Space Sci. Rev.*, this volume.

Burlaga, L. F. and Ogilvie, K. W.: 1973, 'Solar Wind Temperature and Speed', *J. Geophys. Res.* **78**, 2028.

Burnett, D. S. *et al.*: 2003, 'The Genesis Discovery Mission: Return of Solar Matter to Earth', *Space Sci. Rev.*, this volume.

Galvin, A. B.: 1997, 'Minor Ion Composition in CME-Related Solar Wind', in N. Crooker, J. A. Joselyn and J. Feynman (eds.), *Coronal Mass Ejections, Geophysical Monograph 99*, Amer. Geophys. Un., Washington, DC, pp. 253.

Gloeckler, G., Fisk, L. A., Hefti, S., Schwadron, N. A., Zurbuchen. T. H., Ipavich, F. M., Geiss, J., Bochsler, P., and Wimmer-Schweingruber, R. F.: 1999, 'Unusual Composition of the Solar Wind in the 2–3 May 1998 CME Observed with SWICS on ACE', *Geophys. Res. Lett.* **26**, 157.

Gosling, J. T., Pizzo, V., and Bame, S. J.: 1973, 'Anomalously Low Proton Temperatures in the Solar Wind Following Interplanetary Shock Waves: Evidence for Magnetic Bottles?', *J. Geophys. Res.* **78**, 2001.

Gosling, J. T., Bame, S. J., Feldman, W. C., McComas, D. J., Phillips, J. L., and Goldstein, B. E.: 1993, 'Counterstreaming Suprathermal Electron Events Upstream of Corotating Shocks in the Solar Wind beyond 2 AU: Ulysses', *Geophys. Res. Lett.* **20**, 2335.

Gosling, J. T., Birn, J., and Hesse, M.: 1995, 'Three-Dimensional Magnetic Reconnection and the Magnetic Topology of Coronal Mass Ejection Events', *Geophys. Res. Lett.* **22**, 869.

Gosling, J. T., Skoug, R. M., and Feldman, W. C.: 2001, 'Solar Wind Electron Halo Depletions at 90 deg Pitch Angles', *Geophys. Res. Lett.* **28**, 4155.

Ho, G. C., Hamilton, D. C., Gloeckler, G., and Bochsler, P., 2000, 'Enhanced Solar Wind $^3He^{2+}$ Associated with Coronal Mass Ejections', *Geophys. Res. Lett.* **27**, 309.

Lopez, R. E.: 1987, 'Solar Cycle Invariance in the Solar Wind Proton Temperature Relationships', *J. Geophys. Res.* **92**, 11189.

McComas, D. J. *et al.*: 1998, 'Solar Wind Electron Proton Alpha Monitor (SWEPAM) for the Advanced Composition Explorer', *Space Sci. Rev.* **86**, 563.

Neugebauer, M., Goldstein, R., and Goldstein, B. E.: 1997, 'Features Observed in the Trailing Regions of Interplanetary Clouds from Coronal Mass Ejections', *J. Geophys. Res.* **102**, 19.

Richardson, I. G. and Cane, H. V.: 1995, 'Regions of Abnormally Low Proton Temperature in the Solar Wind (1965–1991) and Their Association with ejecta', *J. Geophys. Res.* **100**, 23 397.

von Steiger, R., Schwadron, N. A., Fisk, L. A., Geiss, J., Gloeckler, G., Hefti, S., Wilken, B., Wimmer-Schweingruber, R. F., and Zurbuchen, T. H.: 2000, 'Composition of Quasi-Stationary Solar Wind Flows from Ulysses/Solar Wind Ion Composition Spectrometer', *J. Geophys. Res.* **105**, 27217.

Zwickl, R. D., Asbridge, J. R., Bame, S. J., Feldman, W. C., Gosling, J. T., and Smith, E. J.: 1983, 'Plasma Properties of Driver Gas Following Interplanetary Shocks Observed by ISEE-3', in M. Neugebauer(ed.), *Solar Wind Five; NASA Conference Proceedings 2280*, NASA, Washington, DC, pp. 711.